Monte-Carlo Simulation

Monte-Carlo techniques have increasingly become a key method used in quantitative research. This book introduces engineers and scientists to the basics of using the Monte-Carlo simulation method which is used in Operations Research and other fields to understand the impact of risk and uncertainty in prediction and forecasting models.

Monte-Carlo Simulation: An Introduction for Engineers and Scientists explores several specific applications in addition to illustrating the principles behind the methods. The question of accuracy and efficiency with using the method is addressed thoroughly within each chapter and all program listings are included in the discussion of each application to facilitate further research for the reader using Python programming language.

Beginning engineers and scientists either already in or about to go into industry or commercial and government scientific laboratories will find this book essential. It could also be of interest to undergraduates in engineering science and mathematics, as well as instructors and lecturers who have no prior knowledge of Monte-Carlo simulations.

Monte-Carlo Simulation
An Introduction for Engineers and Scientists

Alan Stevens

CRC Press
Taylor & Francis Group
Boca Raton London New York

CRC Press is an imprint of the
Taylor & Francis Group, an **informa** business

First edition published 2023
by CRC Press
6000 Broken Sound Parkway NW, Suite 300, Boca Raton, FL 33487-2742

and by CRC Press
4 Park Square, Milton Park, Abingdon, Oxon, OX14 4RN

CRC Press is an imprint of Taylor & Francis Group, LLC

© 2023 Taylor & Francis Group, LLC

Library of Congress Cataloging-in-Publication Data

Names: Stevens, Alan (Mathematician), author.
Title: Monte-Carlo simulation : an introduction for engineers and scientists / Alan Stevens.
Description: First edition. | Boca Raton, FL : CRC Press, 2023. | Includes
bibliographical references and index. | Summary: "Monte Carlo techniques have
increasingly become a key method used in quantitative research. This book introduces
engineers and scientists to the basics of using the Monte-Carlo simulation method
which is used in Operations Research and other fields to understand the impact of
risk and uncertainty in prediction and forecasting models. It explores several specific
applications in addition to illustrating the principles behind the methods. Beginning
engineers and scientists either already in or about to go into industry or commercial and
government scientific laboratories will find this book essential"-- Provided by publisher.
Identifiers: LCCN 2022014193 (print) | LCCN 2022014194 (ebook) |
ISBN 9781032280776 (hbk) | ISBN 9781032280806 (pbk) | ISBN 9781003295235 (ebk)
Subjects: LCSH: Quantitative research. | Engineering mathematics. | Monte Carlo method.
Classification: LCC T57.64 .S74 2023 (print) | LCC T57.64 (ebook) |
DDC 518/.282--dc23/eng/20220701
LC record available at https://lccn.loc.gov/2022014193
LC ebook record available at https://lccn.loc.gov/2022014194

ISBN: 978-1-032-28077-6 (hbk)
ISBN: 978-1-032-28080-6 (pbk)
ISBN: 978-1-003-29523-5 (ebk)

DOI: 10.1201/9781003295235

Typeset in Times
by KnowledgeWorks Global Ltd.

Dedication

To Sandra, for times gone by and those still to come.

Contents

Contents   ix

Preface

The purpose of this book is to introduce engineers and scientists to the basic ideas underlying the Monte-Carlo simulation method. It does this in the first instance by exploring a number of specific applications, which you, the reader, are encouraged to try for yourself. In addition to illustrating the principal ideas behind Monte-Carlo simulation these applications generate questions about the accuracy and efficiency of the method, which are also addressed.

Most of the applications considered here are drawn from the world of science and engineering, albeit in simplified form, in order to provide a concrete context, which, it is hoped, will ease your task in assimilating the essential elements of the Monte-Carlo process. The first example – the calculation of the number π in chapter 2 – is drawn from the world of mathematics rather than science or engineering, but is included as a demonstration of the principle of Monte-Carlo simulation that is particularly easy to grasp whatever your background might be.

Computer programs are well-nigh essential for performing Monte-Carlo calculations (one exception is noted in chapter 2), so simple program listings are included in the discussion of each application. The listings are given here using the open-source Python programming language, making use of the NumPy (NUMerical PYthon) library. I have made no attempt to generate the most efficient Python programs, as the focus of this book is on Monte-Carlo simulation, not programming. However, explanatory comments in the listings and the text should make it easy to translate the coding into other programming languages, such as Matlab, Mathcad, Maple, Visual Basic, etc. without too much difficulty.

About the Author

Alan Stevens spent his working life at Rolls-Royce as a mathematical modeller, dealing mainly with design, safety and performance calculations for nuclear reactors. Rolls-Royce designs and manufactures the nuclear reactors that power the Royal Navy's submarines. This included engineering heat transfer and fluid flow, reactor physics and whole plant modelling. In retirement, he has sat on several committees of the Institute of Mathematics and Its Applications (IMA) in the UK, including its Executive Board and governing Council. He spent four years as its VP of Communications.

1 The Basic Idea

1.1 INTRODUCTION

Monte-Carlo simulation can be viewed as 'experimental' calculation, in which we use random numbers to conduct experiments. Typically, we perform the calculations on a computer using anything from hundreds to billions of random numbers. The basic idea is to run a number of trials, or 'experiments', using a mathematical model of the system of interest. For each trial, we select specific values for various input parameters from appropriate statistical distributions. We collect values of the output parameters from all the trials and manipulate these to obtain the measures of interest, like the overall average and its variability.

For example, imagine a manufacturer produces threaded bolts, all nominally the same. In reality each bolt will exhibit small differences in length, major and minor diameters, material properties, etc., which will give rise to small differences in bolt properties, such as effective stiffness. To determine the corresponding variability in stiffness, an engineer might directly measure the stiffnesses of a number of bolts using an appropriate instrument – the experimental approach. Alternatively, if the variabilities of the length, diameters, material properties and so on are known, the engineer might determine the stiffness of a bolt by randomly selecting a value of each of these parameters and calculating the corresponding effective stiffness. By repeating this calculation, a large number of times, using different random selections of each parameter every time, the engineer would again be able to determine the variability in stiffness – the Monte-Carlo simulation approach.

Clearly, for any such simulation to be feasible, not only must a calculational model exist, but the distributions of its input parameters must be known.

The term 'Monte-Carlo', suggested by the random selections on the roulette wheels of the gambling casinos in the city of Monte-Carlo, was introduced by Nicholas Metropolis, John von Neumann and Stanislav Ulam while working on the atomic bomb at Los Alamos during World War II. However, the method, if not the name, was in use much earlier, as we shall see.

Applications of the technique can be split into two broad categories. In the first, input data constructed using random numbers are used to generate purely deterministic output parameters; in the second, they are used to generate probabilistic or statistical output measures. We'll consider examples of both.

The effectiveness of Monte-Carlo simulation depends crucially on the use of good pseudo-random number generators. A brief discussion of a few different sorts of pseudo-random number generators is included in Appendix A, though for the purposes of this book we will assume that those built-in to the software being used are sufficiently good.

DOI: 10.1201/9781003295235-1

The Basic Idea

1.1 INTRODUCTION

2 Buffon's Needle

2.1 BACKGROUND

Historically, the first example of a computation using the Monte-Carlo method arose from a probability question, to do with tossing needles onto a lined surface, addressed by George Louis Leclerc, Comte de Buffon, which he described in 1777 in his treatise 'Essai d'arithmetique morale' [1]. This gives a rather surprising way of using random numbers (representing the position and orientation of the tossed needles) to generate the number π.

Picture a plane surface with parallel lines drawn on it spaced a distance D apart (Figure 2.1). Suppose that a needle, of length L (shorter than D), is tossed randomly onto the surface. Buffon showed that the probability of the needle intersecting one of the lines is $2L/\pi D$. By randomly tossing lots of needles onto such a lined surface, we can estimate this probability by counting the number that crosses a line and dividing by the total number tossed. The resulting fraction, let's call it f, should be roughly equal to $2L/\pi D$, so, with a little algebraic manipulation, we can estimate π from:

$$\pi \approx \frac{2L}{fD} \tag{2.1}$$

We can simulate the process by generating pairs of random numbers. One of each pair represents the distance, y, of the centre of a needle from one of the lines; the other represents the angle, θ, the needle makes with the direction of the lines. If $(L/2)\sin\theta$ is greater than y, the needle crosses the lower line; if it is greater than $D - y$, the needle crosses the higher line; if it is smaller than both y and $D - y$, the needle does not touch either line.

2.2 COMPUTER SIMULATION

By generating random numbers for y taken uniformly between 0 and D, and random numbers for θ taken uniformly between 0 and 180°, we can simulate the process of dropping needles onto a lined surface and count the number of 'hits', i.e., the number that touches or crosses a line. Then, using Equation 2.1, we can estimate the value of π. The Python program in listing 2.1 does this for 1000 'needles'.

In Python, anything preceded on a line by the hash sign, #, is treated as a comment and ignored during the calculations. The first two non-comment lines in listing 2.1 ensure that the required functions are available for use in the subsequent calculations. Some of the functions have the prefix np, indicating that they are from the NumPy library (see [2] for a detailed introduction to the Python language).

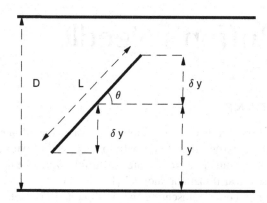

FIGURE 2.1   Buffon's needle.

Listing 2.1 Buffon's needle

```
# Buffon.py
# Estimates pi using Buffon's needle method

import numpy as np
from  numpy.random import rand

# D = spacing, L = needle length, N = number of needles
D, L, N = 10, 8, 1000

# y = needle centres
y = rand(N)*D

# vertical distances between needle centres and needle ends
r = np.zeros(N)
yy = np.zeros(N)
for i in range(N):
    R = 2
    while R>1:
        X = rand()
        Y = rand()
        R = np.sqrt(X**2 + Y**2)
    r[i] = R
    yy[i] = Y
deltay = (L/2)*yy/r

# hits = True if needle touches line or False if not
hits = (deltay>= y) +  (deltay >= D - y)

# f = Fraction that cross a line
f = np.mean(hits)

# Monte-Carlo estimate of pi
PI = 2*L/(f*D)
```

The function, **rand(N)**, generates N random numbers in the range 0–1, so that subsequent lines generate N random values of y and δy. The value of δy is calculated by scattering random points over a unit square, then ignoring those that lie outside a radius of one unit from the origin so we only consider points within the unit quarter-circle. The y-value is then scaled to correspond to a needle half-length of $L/2$. This, in effect, results in a uniform distribution of the angle θ of Figure 2.1. The Boolean expressions, deltay≥y and deltay≥D-y, return **True** or **False** as appropriate, so that **hits** is a vector of true and false values. Because **True** is treated as 1 and **False** as 0, the fraction, f, of needles that touch a line is calculated directly using the mean function. An estimate of π is then calculated using Equation 2.1.

Running this program three times produced the following values (to three decimal places): 3.194, 3.143, 3.036 (if you run this yourself, you will almost certainly come up with different values because of the random numbers involved). Because of the use of random numbers, another run would undoubtedly produce yet another different result. Since we know the true value of π is 3.142 to three decimal places, then clearly, we should not quote the results of our Monte-Carlo simulations this precisely. The above three values only agree with each other and with the true value to one significant figure! How many needles would we need to drop to be confident of achieving greater precision?

For Monte-Carlo simulation in general, we may be interested in either or both of the two related questions: How confident can we be in our result for a given number of trials? How many trials do we need to attain a specific level of confidence in the result? We will consider this more carefully in section 2.4.

2.3 FEATURES OF THE SIMULATION

The importance of this program is not the value we obtain for π of course! It does, however, allow us to identify the important parts of the Monte-Carlo process.

First, and foremost, is the calculational model of the system of interest. In our case, this consists of the mathematical Equation 2.1 that describes the relationship between the parameter of interest, π, and the other parameters, L, D and f, together with the logic connecting f to y and θ. In most real-world situations, the model is likely to be much more complicated than this, possibly consisting of a large computer program in its own right.

Second is the data required by the model. This comprises some fixed data (e.g., the values of L and D) and some data that are described by statistical distributions (e.g., the values of needle position and angle). In our case, the statistical distributions are uniform probability density functions ranged between specified upper and lower limits. In general, we may have non-uniform distributions and not all parameters need to have the same type of distribution.

Next comes the simulation 'wrapper' – the process of looping through the model calculations, selecting different data values from the statistical distributions and recalculating the corresponding parameter (or parameters) of interest. The 'looping' in the Python program shown in listing 2.1 is done partly explicitly using **for** and **while** loops, and partly implicitly using Python's array operations.

2.4 PRECISION

An important part of the model is the determination of the precision associated with the result, or the degree of confidence we have in it. To do this, we need many samples, or trials, in each of which we obtain an estimate of π. We can't really consider the throw of a single needle to be a trial in its own right as each individual needle can only result in an outcome of 0 or 1, neither of which would generate a sensible estimate! Hence, we need to group many needles together. For the program in listing 2.1, we grouped the 1000 needles into a single trial.

One way of getting more estimates is simply to run the program many times, each with a different set of random numbers, recording the resulting estimate each time. For our Buffon program this is a trivial matter and we can repeat the calculation thousands of times with no difficulty.

Figure 2.2 shows the resulting scatter diagram of estimates obtained from 1000 trials. We can see that the estimates are scattered about the true value (shown by the dotted line).

By generating a histogram of the resulting estimates, as in Figure 2.3, we can get a better sense of the underlying variability (the histogram area is normalised to unity). Although uniformly distributed random numbers were used for needle position and angle, the histogram looks roughly normal, as can be seen by the superimposed normal curve.

We use the overall mean, μ, standard deviation, σ, and standard error, se, from the trial average estimates of π together with the central limit theorem (which says that the distribution of sample means approaches a normal distribution as the sample

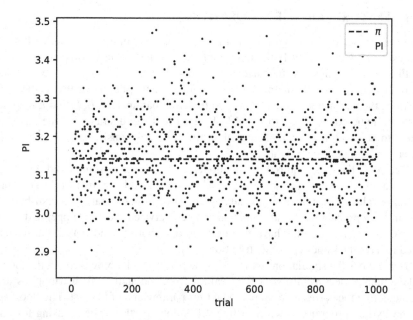

FIGURE 2.2   Monte-Carlo trial estimates of π.

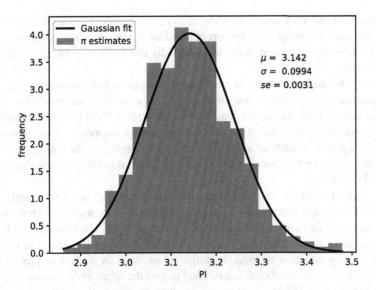

FIGURE 2.3 Histogram of π estimates, with superimposed normal curve.

size tends to infinity) to estimate the likely precision of our overall best estimate. These three measures are defined as follows (where x stands for PI here):

$$\mu(x) = \frac{\sum_{i=1}^{N} x_i}{N}, \ \sigma(x) = \sqrt{\frac{\sum_{i=1}^{N} (x_i - \mu(x))^2}{N-1}}, \ se(x) = \frac{\sigma(x)}{\sqrt{N}}$$

The values of the mean, standard deviation and standard error are shown in Figure 2.3. From these we can determine that 95% of the sample values lie between $\mu - 1.96\sigma = 2.95$ and $\mu + 1.96\sigma = 3.34$, and that our overall best estimate of π lies between $\mu - 1.96se = 3.138$ and $\mu + 1.96se = 3.150$ to 95% confidence. As it happens, we know that these limits do, indeed, bound the true value, though for real-life problems we won't know the true value (or there wouldn't be any point in doing the Monte-Carlo calculations!). The upper and lower limits for our best estimate agree only when rounded to one significant figure, so the most we can really say, based on this set of calculations, is that π is about 3.

The standard error reduces in proportion to the square root of the number of samples, which suggests that to get two significant figures (basically, an improvement of a factor of 10) to the same confidence level we would need to repeat the calculation with 100 times as many trials (and hence needles).

2.5 PHYSICAL SIMULATION

A computer simulation of Buffon's needle is straightforward, but it can also be fun to do a physical simulation. Use, say, cocktail sticks for the needles and ruled sheets of paper or card for the surface with parallel lines. It's relatively slow, of course, and

I'd suggest estimating π after every fifty or so needles, rather than after every needle. In the days before computers, this sort of physical simulation was the only way to do it, and, surprisingly, several individuals not only did it this way but also published their results!

Some published results are summarised in a paper by N.T. Gridgeman [3]. Some of these 'results' are absurd. For example, one appears to be accurate to seven significant figures from the use of just 3408 needles! Not all the numbers of trials are round numbers, raising the suspicion that the authors stopped at their closest estimate. Gridgeman is scathing about all these results, saying of the authors: 'The sole value remaining in their work is its furnishing material to illustrate paralogy, humbug and gullibility'.

Gridgeman is harsh, but we must surely agree that no one in their right mind would seriously use this technique to calculate digits of π, with or without a computer; there are far more direct and efficient ways to do so! The best that can be said of the Buffon's needle approach is that it provides a simple illustration of the basic Monte-Carlo simulation method. It is also an example of a class of problems for which Monte-Carlo simulation can be used to find the areas under curves.

Areas? Curves? Where do they come in? Buffon's needle is all about straight lines, isn't it? However, given the fact that π is involved, perhaps we shouldn't be too surprised to find areas and curves hidden away somewhere. To make them explicit we need to derive Buffon's formula.

2.6   BUFFON'S FORMULA

Picture a graph of y the needle centre, plotted against θ, the needle orientation (Figure 2.4). As before, take the range of y to be from 0 to D and that of θ to be from

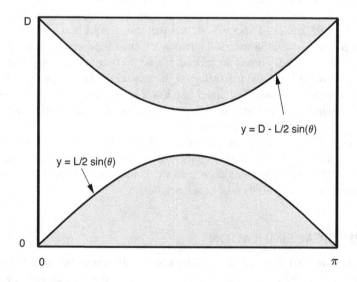

FIGURE 2.4   Buffon's needle plots of y against θ.

0 to 180° π radians). Plot the lines given by $y = L/2\sin\theta$ and $y = D - L/2\sin\theta$. Now we see curves and areas! Referring to Figure 2.4, we can see that the shaded areas represent values of y and θ for which we scored hits in our simulation. Expressed as a fraction of the rectangular area enclosed within the ranges 0 to D (y values) and 0 to π (θ values), they give us the value of f.

Representing the shaded areas by A, we have:

$$f = \frac{A}{\pi D} \tag{2.2}$$

In this case, the shaded area is given by: $A = 2\int_0^\pi \left(\frac{L}{2}\right)\sin\theta\,d\theta$, or $A = 2L$, so that f becomes:

$$f = \frac{2L}{\pi D} \tag{2.3}$$

This is Buffon's formula, which we previously rearranged to give Equation 2.1.

If, instead of needles and lines, we start with Figure 2.4, we can imagine scattering points at random inside the rectangle. The number of points that land within the shaded areas divided by the total number that land within the rectangle gives us the fraction, f, which is related to π through Equation 2.3.

It is this interpretation that supplies the clue to generalising the approach to enable us to calculate the areas bounded by other curves. The right-hand side of Equation 2.2 is the ratio of the area of interest to that of an easy-to-calculate area (in this case a rectangle) in which the area of interest is embedded. Since the area under a curve is just the definite integral of the function describing the curve, this allows us to use the Monte-Carlo simulation method to find the integral of a possibly awkward function for which there is no analytical solution. The next chapter considers a concrete example.

2.7 EXERCISES

2.7.1 Use a Buffon's needle simulation to estimate π for numbers of trials up to 10^7 or more. How rapidly does the accuracy increase?

2.7.2 What happens if L is greater than D? Equation 2.1 is still valid, but you will need to modify the logic of the simulation if you choose L to be greater than $2D$.

2.7.3 Imagine a square of side 2 inscribed with a circle of radius 1. The ratio of the area of the circle to the area of the square is $\pi/4$. Construct a Monte-Carlo simulation program to estimate the value of π from this arrangement.

2.7.4 Picture n bins and n balls. Construct a Monte-Carlo simulation program that randomly scatters the n balls into the bins. Count the number of bins that don't contain any balls, n_0. Plot the ratio n/n_0 as a function of n. To what well-known constant does the ratio seem to be trying for (no, this time it's not π!)?

REFERENCES

1. George Louis Leclerc, Comte de Buffon (1777) 'Essai d'arithmetique morale', Appendix to "Histoire naturelle générale et particulière", 4, 1777.
2. Christian Hill (2020) Learning Scientific Programming with Python. Cambridge University Press.
3. N.T. Gridgeman (1960) Geometric Probability and the Number π, Scripta Mathematica 25 part 3.

3 Areas and Integrals

3.1 TEMPERATURE PROFILE

The temperature, $T(x,t)$, of a semi-infinite solid, subjected to a step change in surface temperature, Ts, is given as a function of distance, x into the solid, and time, t, by the following equation [1]:

$$\frac{T(x,t)-Ts}{T_0 - Ts} = \frac{2}{\sqrt{\pi}} \int_0^{\eta/2} e^{-z^2} dz \qquad (3.1)$$

T_0 is the initial temperature of the solid, and η, in the upper limit of the integral, is related to distance and time through: $\eta = x/\sqrt{\alpha t}$, where α is the thermal diffusivity of the solid.

The behaviour of the temperature within the solid is illustrated as a function of time and distance in Figure 3.1(a). The separate curves collapse onto the single curve shown in Figure 3.1(b) when plotted against η.

The right-hand side of Equation 3.1 is known as the error function and is written as $\mathrm{erf}(\eta/2)$. It has no analytical solution, so let's solve it numerically using Monte-Carlo simulation. Specifically, we'll choose values of x and t such that $\eta = 2.5$. Figure 3.2 shows the curve given by $\left(2/\sqrt{\pi}\right)e^{-z^2}$ plotted over the range from $z = 0$ to $z = \eta/2$. The area under the curve in Figure 3.2 is $\mathrm{erf}(\eta/2)$, the value of the integral we want.

3.2 SIMULATION

Randomly scattering points uniformly over the bounding rectangle shown in Figure 3.2, we can find the fraction, f, that lies beneath the curve. This fraction is approximately equal to the ratio of our, as yet unknown, area to that of the area of the rectangle. The width of the rectangle (i.e., along the z-axis) is $\eta/2$, the height is $2/\sqrt{\pi}$, so the area of the rectangle in this case is simply $(\eta/2)(2/\sqrt{\pi}) = \eta/\sqrt{\pi}$. That means the area we are trying to find is just $f\eta/\sqrt{\pi}$. The Python program shown in listing 3.1 calculates this area.

Listing 3.1 Temperature integral. Hit or Miss method

```
# HitOrMiss.py
# Temperature integral of erf(eta/2). Hit or miss method.

import numpy as np
from  numpy.random import rand
```

```
# eta = x/sqrt(alpha*t), ymax = rectangle height
eta = 2.5
ymax = 2/np.sqrt(np.pi)
# Number of points
N = 1000
# N random values of z and y
z = rand(N)*eta/2
y = rand(N)*ymax

# hits = True if points lie on or below curve
hits = y<=ymax*np.exp(-z**2)
# f = Fraction of points lying on or below curve.
f = np.sum(hits)/N
# Corresponding estimate of the area.
area = f*eta/np.sqrt(np.pi)
```

Again, **hits** is a vector of true and false values, where true indicates trials for which a point was on or below the curve (i.e., it 'hit' the area we are interested in). The resulting value for the area from one particular run of 1000 points was 0.891 compared with the true value of 0.923. The Monte-Carlo value is in the right ball-park, but clearly it hasn't got there yet! Again, we only have one significant figure of accuracy. How repeatable is our result? How many trials do we need to get an extra significant figure? I'll leave you to investigate this.

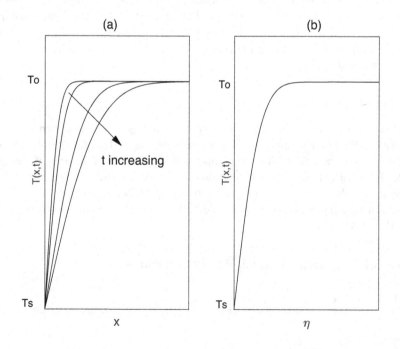

FIGURE 3.1   Temperature vs (a) time and distance (b) η.

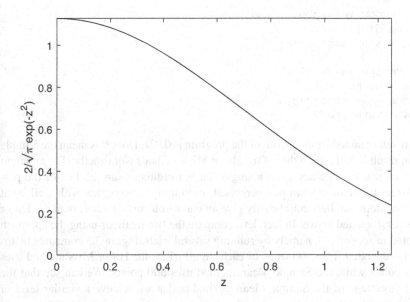

FIGURE 3.2 Curve of the error function integrand.

3.3   SAMPLE MEAN METHOD

For obvious reasons, the Monte-Carlo method we've used above is known as the 'Hit or Miss' method. There is an alternative called the 'Sample Mean' method. The area under a curve is given by the average value of the curve multiplied by the range over which it is integrated. So, if we calculate the value of the function to be integrated for a number of z values taken randomly between 0 and $\eta/2$, take the arithmetic average of these function values and multiply the result by the range of integration (that is $\eta/2$), we get the 'Sample Mean' Monte-Carlo estimate of the integral.

How do the results compare? Listing 3.2 shows the Python program used to calculate the same temperature integral as above, this time using the Sample Mean method, but with a run of only 500 points.

Listing 3.2 Temperature integral. Sample Mean method

```
# SampleMean.py
# Temperature integral of erf(eta/2). Sample mean method.

import numpy as np
from  numpy.random import rand

# eta = x/sqrt(alpha*t), ymax = rectangle height
eta = 2.5
ymax = 2/np.sqrt(np.pi)
# Number of points
```

```
N = 500
# N random values of z and y
z = rand(N)*eta/2
y = ymax*np.exp(-z**2)

 # Average value of y
yave = np.mean(y)
# area = average*range
area = yave*eta/2
```

The result obtained from one run of the program is 0.921. Does this mean the Sample
Mean result is better than that of the Hit or Miss method? Not exactly! The apparent
improvement in accuracy from a single run is fortuitous – an artefact of the particular sequence of random numbers used. Rerunning the program with a different
set of random numbers could easily give an end result further adrift than the Hit or
Miss result quoted above. In fact, let's compare the two methods using the approach
adopted in section 2.4, namely by running several trials. Figure 3.3 compares histo-
grams generated from 500 trials of each. In all trials, the Hit or Miss method uses
1000 points while the Sample Mean method uses 500 points. We can see that the
main advantage of the Sample Mean method is that we achieve a similar level of
accuracy with only half as many random numbers.

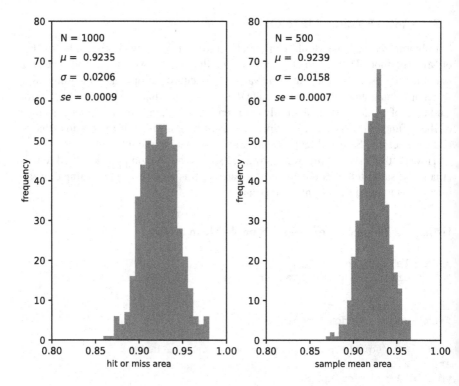

FIGURE 3.3   Temperature integral method comparisons.

3.4 HIGHER DIMENSIONS

In reality, it is rarely a good idea to use a Monte-Carlo method to evaluate simple areas or one-dimensional integrals like those of our examples above. Conventional deterministic techniques (see [2] for example) are computationally much more efficient. However, as the number of dimensions increases, conventional techniques suffer from the 'curse of dimensionality', becoming increasingly inefficient very rapidly. Monte-Carlo techniques do not suffer from this curse and, consequently come into their own for higher dimensional problems.

By 'dimensions' I don't just mean space and time. In the current context, any input from which we intend to perform random sampling counts as a 'dimension'. In this sense, use of the Hit or Miss method in finding the value of the error function above was an example of a two-dimensional Monte-Carlo problem because we randomly sampled for values of y and z. In real scientific and engineering problems there could easily be several tens of inputs, giving rise to high dimensionality problems. However, let's not jump in at the deep end! For the next example, in chapter 4, we'll cautiously increase the number of dimensions to four!

3.5 EXERCISES

3.5.1 Repeat the above process using different sets of random numbers and numbers of trials. How many trials do you need to get an area accurate to two significant figures consistently?

3.5.2 Construct a Monte-Carlo simulation program to calculate the area under the curve of the function $y = \sqrt{x}/(1 + 0.1\sin \pi x)$ between the limits $x = 0$ and $x = 1$ (the area is 0.626, correct to three significant figures).

REFERENCES

1. Adrian Bejan (1993) Heat Transfer. John Wiley & Sons, Inc.
2. Richard L. Burden, J. Douglas Faires (2001) Numerical Analysis, 7th Edition. Brooks Cole.

4 Thermal Radiation

4.1 RADIATION VIEW FACTOR

Any surface radiates thermal power in proportion to the fourth power of its absolute temperature. A nearby surface may intercept some of this radiation. The amount intercepted depends on the relative geometrical orientation of the two surfaces. A parameter called the view factor provides the measure of the fraction of radiant energy leaving one surface that arrives at the other (the view factor is also known by other names, such as shape factor, angle factor and configuration factor). Let's calculate the view factor for a specific geometry; namely, that in which radiation transmitted from a rectangular horizontal surface is intercepted by a rectangular vertical surface with which it shares a common edge, as illustrated in Figure 4.1.

Mathematically, the view factor involves a double integral over the two surface areas. The view factor, f_{12}, from the horizontal to the vertical surface (refer to Figure 4.1) is defined explicitly by (see [1] for example):

$$f_{12} = \frac{1}{A_1} \iint \frac{\cos\theta\cos\psi}{\pi r^2} dA_1\, dA_2 \qquad (4.1)$$

The angles θ and ψ are taken between the normals to their respective surfaces and the line, of length r, that joins the infinitesimal areas at each end of the line. Because each integral is over an area, it is itself, in effect, a double integral, making the whole thing a quadruple integral! In spite of this, because of the very simple geometry we are considering here, it is possible to obtain the following exact, analytical solution, [1]. With X, Y and Z as in Figure 4.1, set $a = Z/Y$, $b = X/Y$, $c = a\wedge 2 + b \wedge 2$, then:

$$f_{12} = \frac{1}{\pi b}\left\{\left[\ln\left(\frac{(1+a^2)(1+b^2)}{1+c}\right) + b^2\ln\left(\frac{b^2(1+c)}{c(1+b^2)}\right) + a^2\ln\left(\frac{a^2(1+c)}{c(1+a^2)}\right)\right] \times \right.$$
$$\left. \frac{1}{4} + b\tan^{-1}\frac{1}{b} + a\tan^{-1}\frac{1}{a} - \sqrt{c}\tan^{-1}\frac{1}{\sqrt{c}}\right\} \qquad (4.2)$$

For other geometries where it is possible to carry out the integrals analytically the resulting expressions are often large and unwieldy, like that in Equation 4.2. In general, for more complicated geometries, no analytical solutions are possible, so numerical solutions must be sought. We will avoid any analytical complexities by calculating the view factor using a Monte-Carlo simulation approach.

4.2 SIMULATION

Instead of using a Sample Mean approach, with the formula in Equation 4.1, we will adopt a Hit or Miss approach (almost literally!) to tackle this. We randomly scatter points, representing sources of thermal radiation, all over the horizontal surface,

DOI: 10.1201/9781003295235-4

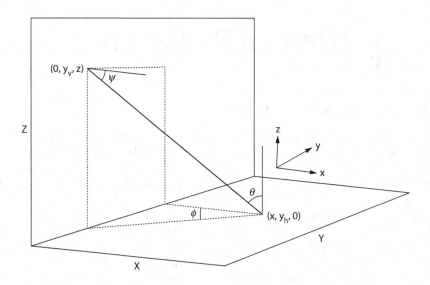

FIGURE 4.1   Radiation from horizontal to vertical surface.

with rays leaving those points isotropically at random angles, and determine which of those rays strike the vertical surface. The view factor is then given directly by the fraction of all the rays emitted that are intercepted by the vertical surface.

The essential steps are detailed in listing 4.1 for the Python program that does the calculations. It uses a horizontal surface of size $X = 1$, $Y = 1$, and a vertical surface of height $Z = 2$, with $N = 1000$ points. The x, y coordinates of the points on the horizontal surface, and the azimuthal angles, ϕ, are chosen with uniformly distributed random numbers. However, the polar angles, θ, are chosen from weighted random numbers. To ensure that each element of solid angle, of the sphere surrounding a source point, receives the same contribution from its source on average, we need to choose θ using $\cos^{-1}(1 - 2u)$, where u is a uniformly distributed random number between 0 and 1. Since this selects angles in the range 0 to π and we just want angles in the range 0 to $\pi/2$, we halve the resulting angles. However, this means we make use of angles that would otherwise miss the vertical surface, so we also divide the view factor by 2 to compensate.

Listing 4.1 Thermal radiation view factors

```
# Viewfactor.py
# Estimates thermal radiation view factor between horizontal
# and vertical touching surfaces.

import numpy as np
from  numpy.random import rand

# Dimensions of surfaces (lengths)
X, Y, Z = 1, 1, 2
```

```
# Number of trials, points
trials, N = 400, 1000
# Create storage space for view factors then loop through
trials
vf = np.zeros(trials)

for trial in range(0, trials):
    # random values on horizontal surface
    x = rand(N)*X
    yh = rand(N)*Y
    theta = np.arccos(1-2*rand(N))/2
    phi = rand(N)*np.pi-np.pi/2
    # points on vertical plane
    dy = x*np.tan(phi)
    yv = yh + dy
    z = np.sqrt(x**2 + dy**2)*np.tan(np.pi/2 - theta)
    # number of points that hit vertical surface
    hits = sum((z>0)*(z<Z)*(yv>0)*(yv<Y))
    # view factors
    vf[trial] = hits/(2*N)

# Summary statistics
mu = np.mean(vf)
sigma = np.std(vf)
se = sigma/np.sqrt(trials)
```

Note that for this problem each set of N points constitutes a single trial; so to get a better final view factor estimate the program carries out 400 trials. The program uses the inbuilt NumPy function, zeros, to create a vector with N elements, all initially zero, in which to store the view factors. The view factors calculated for these trials are illustrated as a scatter diagram in Figure 4.2.

The dashed horizontal line in Figure 4.2 shows the analytically derived view factor of 0.2329 (from the expression given in Equation 4.2). The overall average view factor estimated by the program for the run illustrated is 0.2328, which is in reasonable agreement with the analytical value. However, is this just luck? Would we get the same agreement with a different set of random numbers or a different number of trials? These are important questions – we don't want to be accused of stopping the simulation at a conveniently accurate stage! In real scientific and engineering problems, of course, we wouldn't have an analytical value for comparison, but we would still need to know the precision of our result.

4.3   PRECISION

It is clear from Figure 4.2 that, although the estimated view factors cluster about the true value, there is a significant amount of scatter. We can get a better feel for the nature of the scatter by plotting a histogram of the values. The result of doing this, together with a superimposed normal curve, is shown in Figure 4.3 (the normal curve is rescaled to match the area of the histogram).

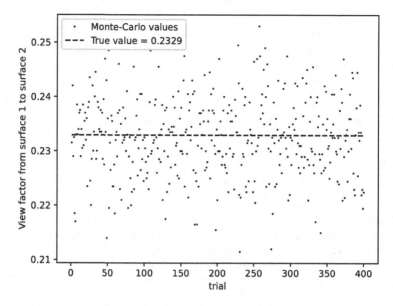

FIGURE 4.2   Monte-Carlo estimates of view factor.

We can now calculate the appropriate statistics and use the standard error on the mean to determine the lower and upper 95% limits. These occur at $\mu - 1.96se = 0.2320$ and $\mu + 1.96se = 0.2336$, respectively. These agree to two significant figures, so the best we can reasonably say (with 95% confidence) on the basis of the present calculation is that the view factor is about 0.23.

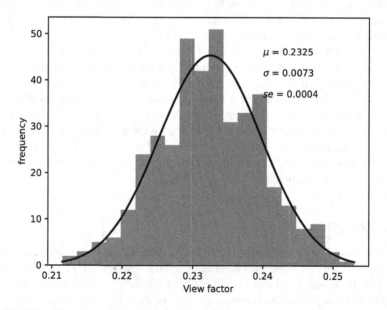

FIGURE 4.3   Histogram of view factor estimates.

4.4 EXERCISES

4.4.1 Calculate the view factor from horizontal to vertical when the two surfaces are infinitely wide (i.e., $Y = \infty$ in Figure 4.1). (The true value is $(1/2)\left(1 + Z/X - \sqrt{\left(1 + (Z/X)^{\wedge}2\right)}\right)$ using the notation of Figure 4.1.)

4.4.2 Calculate the view factor between two infinitely wide surfaces of length, X, joined along one edge, but at an acute angle, θ, to each other. (The true value is $1 - \sin(\theta/2)$.)

REFERENCE

1. Warren M. Rohsenow, James P. Hartnett (1973) Handbook of Heat Transfer. McGraw-Hill.

5 Bending Beams

5.1 NEUTRAL AXIS OFFSET

A beam, curved because of bending forces, experiences tensile stresses in its outer region, away from the centre of curvature, and compressive stresses in the inner region, towards the centre of curvature. The line separating these two regions is called the neutral axis. For a bent beam, the neutral axis does not coincide with the centroidal axis (see Figure 5.1). The greater the curvature, the greater the separation, h, between these two axes. Let's use a Monte-Carlo approach to calculate this separation for a circular cross-section beam.

The distance, s (see Figure 5.2), between the centroidal axis and any point on the beam cross-section, measured from the centre of bending, is used to define the separation, h, through the following ratio of integrals [1]:

$$h = \frac{\int s/(R_c - s)dA}{\int 1/(R_c - s)dA} \tag{5.1}$$

where the integrals are taken over the whole cross-sectional area. R_c is the distance from the centre of curvature of the centroidal axis, R_n is the distance of the neutral axis and R is the radius of the circle.

Again, because of the simple geometry, an analytical solution is possible [1]:

$$h = R_c - \frac{R^2}{2\left(R_c - \sqrt{R_c^2 - R^2}\right)} \tag{5.2}$$

5.2 SIMULATION

By scattering points randomly over the circular cross-section, calculating the integrands for each point, averaging for all points and taking the ratio of the two averages we should obtain a sample mean, Monte-Carlo estimate of h. Figure 5.2 shows how s, a random point, (r, θ) and the centroidal axis are related. Quantitatively, $s = r\sin\theta$.

Because the cross-section is circular, it seems sensible to choose random radii and angles, rather than random Cartesian coordinates, for the scattered points. However, some care is required in doing this. Figure 5.3(a) shows the distribution of 500 points on a circle, where both the radii and the angles have been chosen from uniform random distributions between 0 and R (the radius of the circle) for the radii, and 0 and 2π for the angles. Clearly the points are not uniformly distributed over the circle – there is a greater density towards the centre. This is because for a larger

DOI: 10.1201/9781003295235-5

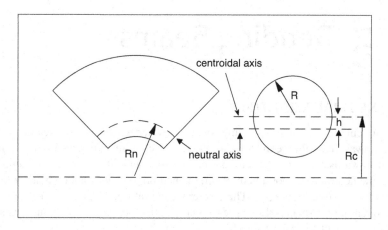

FIGURE 5.1   Curved beam.

radius the area available over which to spread the points is larger. To maintain a uniform average density, we need to select area uniformly. Since area is proportional to the square of the radius, this means selecting radius-squared uniformly, or, equivalently, selecting radii from the square roots of uniform random numbers. This has been done in Figure 5.3(b), where the points are seen to have a much more uniform spread.

With this in mind, listing 5.1 shows a Python program which may be used to generate the centroidal-neutral axis separation for the case of $R_c/R = 1.5$. The function, **trapz**, uses the trapezoidal rule to do the integrations. The estimate of the neutral

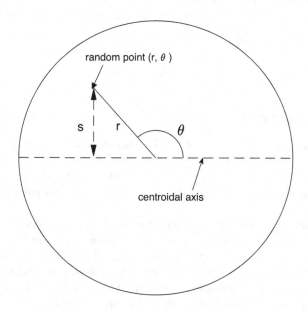

FIGURE 5.2   Beam cross-section.

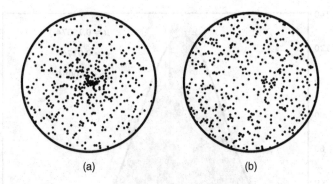

FIGURE 5.3 Random points in a circle (a) radii (b) radii^2 chosen randomly.

axis offset using 1000 points was 0.198, around 3.7% higher than the true value, 0.191, obtained from Equation 5.2.

Listing 5.1 Neutral axis offset

```
# CurvedBeam.py
# Estimates distance, h, between centroidal and neutral axes
# of a curved beam of circular cross-section

import numpy as np
from numpy.random import rand

# R, Rc = Radius of circle and centroidal axis respectively
R, Rc = 1, 1.5
# Number of points
N = 1000
# r = radii, theta = angles, s = displacements from axis
r = np.sqrt(rand(N)*R**2)
theta = rand(N)*2*np.pi
s = r*np.sin(theta)
# integrand denominator and numerator
den = 1/(Rc - s)
num = s*den
# Ratio of integrals gives h, the neutral axis offset
h = np.trapz(num)/np.trapz(den)
```

5.3 PRECISION

Again, we repeat the calculation shown in listing 5.1, 1000 times. Figure 5.4 displays a histogram of the results together with a superimposed normal curve. The 95% upper and lower limits of the standard error on the mean are 0.192 and 0.190, respectively. These agree to two significant figures, so we can say that the neutral axis offset is about 0.19.

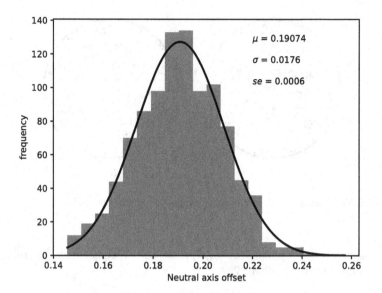

FIGURE 5.4   Curved beam neutral axis offset histogram.

5.4   EXERCISES

5.4.1 Another way of generating random points uniformly over a circle of radius, R, is to calculate x and y coordinate values uniformly between $-R$ and R, but then reject those points for which $\sqrt{x^2 + y^2}$ is greater than R. Repeat the calculation of the separation, h, above, using this method of generating the random points.

5.4.2 How does the neutral axis offset, h, vary with the ratio R_c/R? Try a range of ratios from, say, $R_c/R = 1$ to $R_c/R = 5$.

REFERENCE

1. Joseph E. Shigley, Charles R. Mischke, Richard G. Budynas (2004) Mechanical Engineering Design, 7th Edition. McGraw-Hill.

6 Torus Segment

6.1 VOLUME OF A SEGMENT OF TORUS

In the previous two problems, we had the choice of using either the Hit or Miss or the Sample Mean method of Monte-Carlo simulation. This was possible because of the existence of an explicit integral expression for the quantity of interest in both cases. Sometimes, such an explicit expression is difficult to obtain, but the parameters of interest can still be simply stated in terms of bounding constraints. For this type of scenario, the Hit or Miss method is easier to implement.

For example, imagine we wish to find the volume of a segment of a torus, where the segment has been created by slicing through a torus, as shown in Figure 6.1. The slice is offset from a central plane of symmetry, so that the segment is smaller than half the full torus. Specifically, we'll assume the 'hole' in the torus has radius two units, its 'tube' has an internal radius of one unit and the slicing occurs one unit from a central plane of symmetry.

6.2 SIMULATION

One approach, using the Hit or Miss method, would be to enclose the torus segment within a rectangular box that has sides comprising the x-y planes at z-values of 1 and -1, the x-z planes at y-values of 4 and -4 and the y-z planes at x-values of 1 and 4. By randomly scattering points within the rectangular box (which has the easily calculated volume of 48 cubic units) and determining the fraction that falls within the region occupied by the torus segment, we could obtain an estimate of the volume of the latter.

How do we determine if a random point within the rectangle lies within the torus? Look at an r-z slice through the torus, where the radius, r, is the vector from the origin to the point set by the random values of x and y (see Figure 6.2). If R_c is the radial distance from the centre of the 'hole' to the centreline of the torus (i.e., the sum of the hole and tube radii), R_t is the radius of the 'tube', and z is the random value of the z-coordinate, then the point will lie within the torus segment if $\sqrt{(r - R_c)^2 + z^2} \leq R_t$. The value of r is just $\sqrt{x^2 + y^2}$ of course. (In our case, we have $R_t = 1$ and $R_c = 3$).

However, to do this exactly as suggested above would be very inefficient, as the volume of a large part of the segment may be found as a simple fraction of the volume of a complete torus. Indeed, the volume of the sector between θ_{min} and θ_{max} (see Figure 6.3) is simply the fraction $2\theta_{min}/2\pi$ of the volume of the whole torus. It is just the regions near the ends, between θ_{min} and θ_{max} for example, that are awkward to calculate, so we should focus the Monte-Carlo calculations on those.

The angle, θ_{min}, is fixed by the radial slice that just clips the inside of the torus segment, and θ_{max} is fixed by the radial slice that just clips the outside. They are given here by $\theta_{min} = \cos^{-1}(0.5)$ and $\theta_{max} = \cos^{-1}(0.25)$. We home in on the simple wedge-shaped

DOI: 10.1201/9781003295235-6

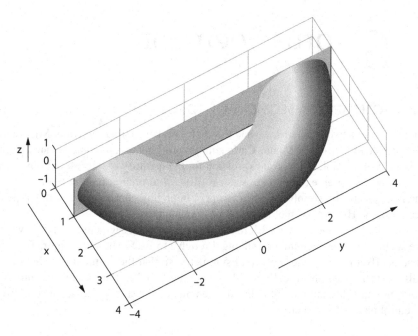

FIGURE 6.1   Segment of torus.

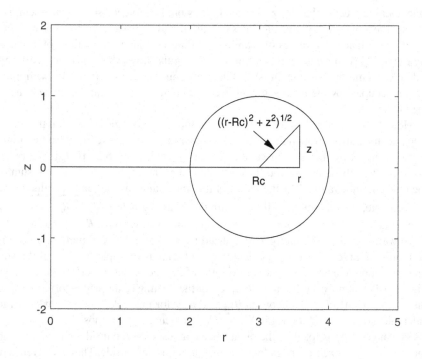

FIGURE 6.2   *r-z* slice through torus segment.

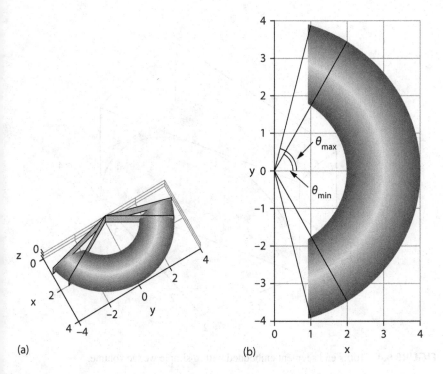

FIGURE 6.3   Torus segment with radial slices marked; (a) general (b) from above.

volume formed from the annular gap between two part-cylinders of radius $R_c - R_t$ and $R_c + R_t$, extending from $z = 0$ to $z = R_t$, and contained between angles θ_{min} and θ_{max} (symmetry above and below $z = 0$ means we don't need to include the negative z values explicitly) – see Figure 6.4.

We can calculate the volume of the torus segment, V_{end}, contained in this wedge using the Hit or Miss method. The overall segment volume is then given by four times V_{end} added to the fraction $2\theta_{min}/2\pi$ of the volume of the whole torus. The volume of the simple wedge is just $\pi\left(\left(R_c + R_t\right)^2 - \left(R_c - R_t\right)^2\right)R_t\left(\theta_{max} - \theta_{min}\right)/2\pi$ (=1.6255 here). The volume of the whole torus is $\pi R_t^2 2\pi R_c$ (= $6\pi^2$ here).

Listing 6.1 shows the Python program used to do the calculation. Notice that, as in the previous chapter, because we use $r - \theta$ geometry within the wedge, we must generate a uniformly random distribution of the square of the radius to ensure the points are not biased towards the inner regions of the wedge. There are two conditions that must be satisfied simultaneously for the random point to 'hit' the torus segment within the wedge. The first is that $\sqrt{\left(r - R_c\right)^2 + z^2} \leq R_t$, as noted above; the second is that $r\cos\theta \geq 1$ (if the latter isn't true the point will lie behind the y-z slice at $x = 1$).

An example single run of the program produced an estimate of the torus segment volume of 23.029. However, as in previous chapters, we need to calculate many more values.

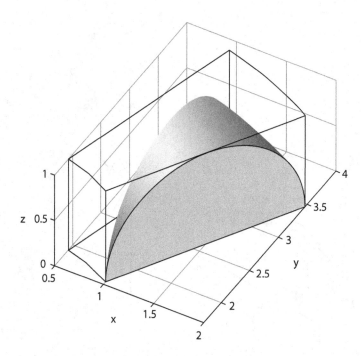

FIGURE 6.4   Torus end segment embedded within simple wedge volume.

Listing 6.1 Torus segment volume

```
# Torussegment.py
# Estimates volume of torus segment from hit-or-miss method
# by scattering random points in wedge bounding a positive
# end-piece of the torus segment.

import numpy as np
from  numpy.random import rand

# Radii of torus centreline and torus "tube"
Rc, Rt = 3, 1
# Inner and outer radii and their squares
Rmin, Rmax = Rc - Rt, Rc + Rt
Rminsq, Rmaxsq = Rmin**2, Rmax**2
# Bounding angles and heights
thetamin, thetamax = np.arccos(1/2), np.arccos(1/4)
zmin, zmax = 0, Rt
# Volume of wedge
Vwedge = 0.5*(thetamax - thetamin)*(Rmax**2 - Rmin**2)*(zmax
- zmin)
# Volume of torus segment between +/- thetamin
Vseg = (thetamin/np.pi)*2*np.pi*Rc*np.pi*Rt**2
# Number of points
```

```
N = 1000

# Generate N random values of radius, angle and height
r = np.sqrt(Rminsq + rand(N)*(Rmaxsq - Rminsq))
theta = thetamin + rand(N)*(thetamax - thetamin)
z = zmin + rand(N)*(zmax - zmin)

# Calculate if the points hit the torus piece or not
hits = ( np.sqrt( (r-Rc)**2 + z**2 )<=Rt ) * ( r*np.
cos(theta)>=1 )

# Fraction of points within the torus piece
f = np.mean(hits)

# Estimate of torus segment volume
V = 4*f*Vwedge + Vseg
```

6.3 PRECISION

Once again, with the program of listing 6.1 representing a single trial, we repeat
it for 1000 trials. Figure 6.5 shows a histogram of the resulting volumes, with a
superimposed normal distribution. The overall mean volume of the torus segment
is seen, in Figure 6.5, to be about 23.18. The upper and lower 95% confidence
limits on the mean are 23.19 and 23.18, respectively, so we can say the volume is
approximately 23.2.

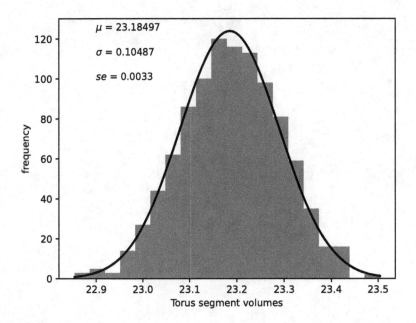

FIGURE 6.5   Torus segment volume histogram.

6.4 EXERCISES

6.4.1 Find the volume of a segment of the same torus, where the segment is offset from the central plane of symmetry by *two* units.

6.4.2 How does the method used in listing 6.1 compare with the straightforward approach of enclosing the torus segment in a rectangular box that has sides comprising the x-y planes at z-values of 1 and -1, the x-z planes at y-values of 4 and -4 and the y-z planes at x-values of 1 and 4, within which the random points are scattered?

7 Radiation Shielding

7.1 DIFFUSION

Diffusion (of pollen grains, neutrons, chemicals, etc.) arises from random collisions of particles. We are normally interested in the macroscopic characteristics of diffusing substances (e.g., flux, concentration) and hence, traditionally, derive continuum equations to describe average properties of large numbers of colliding particles (see [1] for example).

However, we may also adopt a microscopic perspective and explicitly follow individual particles as they take a random walk of multiple collisions. Random numbers are used to generate factors such as the distance between collisions and the direction of travel after a collision. By generating a large number of such random walks, we can form averages, and other statistics, of the characteristics of interest. This, of course, is a Monte-Carlo process.

By way of example, let's consider the transmission of gamma rays through a radiation shielding material.

7.2 GAMMA-RAY SHIELDING

Imagine a mono-directional beam of gamma rays of energy, E_0 incident upon a slab of lead of thickness, L. By how much is the exposure to gamma rays reduced by the presence of the slab of lead? This would be easy to answer if, every time a gamma-ray photon interacted with matter, it disappeared. Then the intensity of radiation, I, would decrease exponentially through the slab: $I = I_0 e^{-\mu x}$, where I_0 is the initial intensity, x is distance through the slab and μ, with dimensions of inverse length, is the linear attenuation coefficient and depends on both the material of the slab and the energy of the gamma rays.

For a very thin slab, this expression can, indeed, be used to calculate the intensity of gamma radiation that will get through the slab. However, gamma rays don't always disappear when they interact with matter; they are often simply scattered in a different direction. These scattered rays can in turn undergo possible multiple scatterings and still emerge on the other side of the slab, albeit with reduced energy. In general, for slabs that are not thin, the extra exposure arising from these scattered gamma rays must be taken into account.

One of three events occurs when a gamma photon interacts with an atom. It can undergo pair production, in which it disappears with the creation of an electron-positron pair. We will ignore this as it happens at higher energies than we will consider here. It can initiate the photoelectric effect, in which it disappears with the ejection of an atomic electron. Also, it can undergo elastic scattering with one of the atom's electrons, in which it loses some of its energy (though the overall energy of the interaction is conserved, of course) but does not disappear. This last type is known as Compton scattering.

DOI: 10.1201/9781003295235-7

The photoelectric effect and Compton scattering each have their own linear attenuation coefficient, μ_{pe} and μ_C respectively. The total linear attenuation coefficient, μ_T, is simply the sum of these. The ratio, $\mu_{pe}/\mu_T = 1 - \mu_C/\mu_T$, gives the probability that the interaction results in the photon being absorbed. The inverse of μ_T measures the mean free path travelled by a gamma photon between interactions.

Unfortunately, the linear attenuation coefficients for both the photoelectric effect and Compton scattering are functions of photon energy (see [2]), and we must take that into account in our simulation. We'll do this by using the following simplified expressions for the total and Compton attenuation coefficients in lead:

$$\mu_T = 0.34\left(1 + \frac{1}{E} + \frac{0.25}{E^{2.8}}\right), \ \mu_C = 0.57(1 - 0.49\ln E) \tag{7.1}$$

With energy, E, in MeV, the values of μ_T and μ_C are in cm^{-1}. We'll assume these are good for the range of energies from 1 down to 0 MeV, though, in reality, they are only a rough fit in the range 1 to 0.1 MeV and are increasingly inaccurate below 0.1 MeV. (Note that in the Python program, the function, **log**, returns the natural logarithm; that is, log to the base e.)

7.3   BUILD-UP FACTOR AND ENERGY DISTRIBUTION

More gamma photons emerge from the far side of the slab than expected from a simple exponential decrease. They emerge with a range of different energies dependent on the number of scatterings they have undergone before they emerge. The ratio of the energy weighted number that emerges to that which would emerge on the basis of an exponential decrease is known as the build-up factor. We can calculate the build-up factor for a particular thickness of lead slab and energy of incident gamma photons, using a Monte-Carlo approach. We can also use the energies carried by the emerging photons to generate a spectrum of the energy distribution of the gamma rays that get through.

7.4   SIMULATION

To be specific, let's assume we have a beam of 1 MeV gamma rays incident on a slab that is two mean free paths thick. That is, the slab is $2/\mu_T$ thick, where μ_T is evaluated at 1 MeV. In this case, we have, using Equation 7.1, a slab that is 2.61 cm thick.

We'll follow each gamma photon as it progresses through the lead slab until it gets absorbed (photoelectric effect), it returns to where it started from, or it gets through to the far side. We'll adopt an event-based approach. An 'event' is an interaction of the photon with an atom, resulting in either absorption or scattering.

At each step of the calculation, the photon will travel from one event to another. The actual distance travelled, r, is determined by multiplying the mean free path, $1/\mu_T$, by a random number. However, unlike our previous examples, this random number is not a uniformly generated random number. Because the loss of photons along the path of travel is essentially that of an exponential decay, the random number in question must be selected from an exponential distribution. This is done by setting it equal to $-\ln u$, where u is a uniform random number between zero and one. That is, $r = -\ln u/\mu_T$.

We decide what type of event occurs by comparing another random number (this time a uniformly distributed one) with the value of $1 - \mu_C/\mu_T$. If the random number is smaller than $1 - \mu_C/\mu_T$ the event is an absorption, otherwise it is a scattering.

If scattering occurs, we need to calculate the new direction of travel and update the photon's energy. The proper way of calculating the Compton scattering of a photon by an atomic electron involves the relativistic Klein-Nishina equation [3]. However, that level of complication is not warranted for our purposes. We will simply assume the photon is randomly scattered in the forward direction (forward relative to its pre-scattering direction that is).

To keep track of the direction of travel we use unit length vectors. The new direction of travel is calculated in three stages (see Figure 7.1). Firstly, a completely random unit vector is calculated. This may point in any direction. Secondly, this vector is added to the unit vector representing the pre-scattering direction (this ensures there is no back-scattering). Lastly, the resulting vector is re-scaled to have unit length by dividing its x-, y- and z-components by the square root of the sum of squares of these components (this is done using the Python function, **norm**, in the program below).

We are interested in how far through the slab the photon has travelled, so, taking the slab through-thickness direction to be the x-direction, we determine the change in the through-thickness distance, Δx, by multiplying the distance of travel, r, by

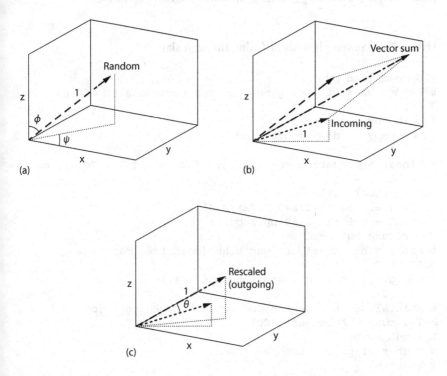

FIGURE 7.1   (a) Random unit vector (b) incoming + random unit (c) outgoing unit vector.

the x-component of the direction vector, u_{vec} or $\Delta x = r \times u_{vec}$. The slab is effectively infinite in the y- and z-directions, so there is no need to track the actual distance travelled in these directions.

To calculate the energy loss in the scattering process, we need the cosine of the angle between the incoming and outgoing photons. To do this, we make use of the fact that the dot product of two vectors, u and v, is given by $u.v = uv\cos\theta$, where θ is the angle between them. With u and v as vectors, the dot product is obtained by multiplying u by v appropriately. With unit length vectors the absolute values multiplying $\cos\theta$ are both unity, so $\cos\theta$ is given directly by the dot product.

The outgoing photon energy, in MeV, is then obtained from (see [2]):

$$E_{out} = \frac{0.511}{1 - \cos\theta + 0.511/E_{in}}$$

Listing 7.1 shows the Python program that performs the Monte-Carlo calculations. It follows 10^6 gamma photons in order to get a large enough number emerging to construct a reasonably smooth energy histogram. The emerging energies are stored in a vector called **energy** (0 is stored if a photon doesn't get through). The ratio of the total transmitted energy to that that would get through if there were no extra energy from emerging scattered gammas, is known as the build-up factor.

Listing 7.1 Gamma photons diffusing through slab

```
# GammaShield.py
# Random walk of 1MeV gamma photons through slab of lead.

import numpy as np
from numpy.random import rand

# Linear attenuation coefficients [cm^-1]. muT = Total, muC =
Compton
def AttnCoeffs(E):
    muT = 0.34*(1+1/E+0.25/E**2.8)
    muC = 0.57*(1-0.49*np.log(E))
    return muT, muC
# muL = "Thickness" of lead slab [mean free paths]
muL = 2.0
# Energy of incident gamma photons [MeV]
E0 = 1.0
# Initialise attn coeffs and actual thickness L [cm])
muT0, muC0 = AttnCoeffs(E0)
L = muL/muT0;
# Number of gamma photons
N = 10**6
 # Create storage space for exit energies
```

```
energy = np.zeros(N)

# Loop through N photons
for p in range(N):

    # Initial Energy and attn coeffs
    E, muT, muC = E0, muT0, muC0
    # Initial direction coordinates (uvec) and step size (x)
    uvec = np.array([1, 0, 0])
    x = -np.log(rand())/muT0

    # Follow the photon's random walk
    keepgoing = True    # Flag to indicate if photon is still
inside slab
    while keepgoing:
        uvec_old = uvec
        if x>L:                      # Photon has passed
through slab
            energy[p] = E
            keepgoing = False
        elif x<0:                    # Photon has returned to
start
            keepgoing = False
        elif rand()<1-muC/muT:       # Photon has been absorbed
            keepgoing = False
        else:                        # Photon has been scattered
            # random angles
            psi = 2*np.pi*rand()
            cphi = 1-2*rand()
            phi = np.arccos(cphi)
            sphi = np.sin(phi)
            # Update unit direction vector
            uvec_rnd = np.array([sphi*np.cos(psi), sphi*np.
sin(psi), cphi])
            uvec = uvec_old + uvec_rnd
            uvec = uvec/np.linalg.norm(uvec)
            # cos(theta)
            ctheta = uvec_old @ uvec
            # Update scattered photon energy, attn coeffs and
distances
            E = 0.511/(1-ctheta+0.511/E)
            muT, muC = AttnCoeffs(E)
            r = -np.log(rand())/muT
            x = x + r*uvec[0]
        # end of if x>L
    # end of while keepgoing loop

# end of for p in range(N) loop

# Build-up factor
BF = np.sum(energy)/(E0*N*np.exp(-muL))
```

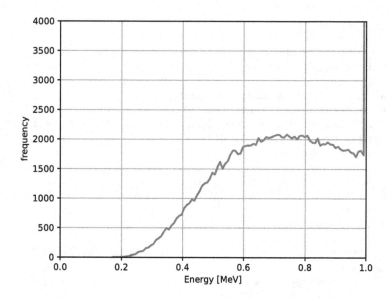

FIGURE 7.2   Gamma photon energy spectrum from lead slab two mean free paths thick.

Using only the non-zero energies – i.e., those of the gamma rays that get through the slab – we can plot a spectrum of the emerging energies, as illustrated in Figure 7.2. The frequency axis has been truncated in Figure 7.2 in order to show the trend of the lower energies more clearly, rather than have the whole curve dominated by the number of photons that travel through the slab without being scattered (the peak value is approximately 3½ times larger than the maximum value seen in Figure 7.2).

7.5   BUILD-UP FACTORS

The build-up factor, calculated from $\Sigma_i\, energy_i/(E_0 N e^{-\mu_T L})$, for one run of the program was 1.688, which compares well with the value of 1.68 tabulated in [2]. We can use our program to calculate values for a range of thicknesses. The resulting values are shown in Figure 7.3. Of course, the usual question, of how repeatable are the results, should still be asked; especially as, for thicker slabs, fewer photons get through, so, for a given total number of input photons, the fluctuations in output from run to run will be greater. However, I'll leave the answer to that as an exercise for you (see exercise 7.6.1).

7.6   EXERCISES

7.6.1 As the slab thickness increases the fraction of photons that get through reduces, which means we should send more through to improve the output statistics. Repeat the build-up factor calculation for a slab thickness of eight mean free paths with 10^7 or 10^8 photons. Is there a noticeable difference?

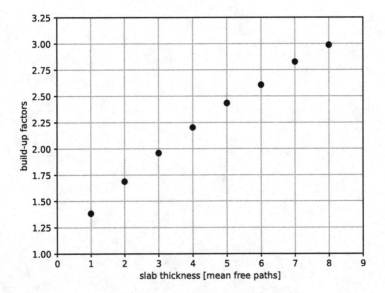

FIGURE 7.3 Build-up factors for mono-energetic beam of gamma rays incident on a lead slab.

7.6.2 A smoke particle diffuses through a room due to isotropic, random collisions with air molecules. In spherical polar coordinates a volume element, dV, is given by $dV = r^2 \sin\theta d\theta d\phi dr = -r^2 d(\cos\theta)d\phi dr$, where r is the distance travelled between collisions, and θ and ϕ lie between 0 and π and 0 and 2π, respectively. This means $\cos\theta$ must be chosen uniformly at random between -1 and $+1$, and ϕ between 0 and 2π. Develop a Monte-Carlo simulation program to calculate the average radial distance (in mean free paths), travelled by the smoke particle, from an arbitrary starting point, as a function of the number of collisions it experiences. For simplicity, assume the distance travelled between each collision is exactly one mean free path.

REFERENCES

1. F. Reif (1965) Fundamentals of Statistical and Thermal Physics. McGraw-Hill.
2. John R. Lamarsh (1975) Introduction to Nuclear Engineering. Addison- Wesley
3. W.E. Burcham (1967) Nuclear Physics: An Introduction. Longmans.

8 Stressed Cylinder

8.1 BUCKLING PROBABILITY

The previous problems in this book have all been concerned with the use of Monte-Carlo simulation to calculate deterministic parameters – areas, volumes, view factors, etc. We now give an example of its use in calculating a probabilistic parameter; namely, a probability of failure – specifically, the probability that a cylinder, subjected to a compressive load, will buckle.

Imagine a thin-walled, hollow cylinder of length, L, internal radius, r, and wall thickness, t, that experiences a compressive axial load, P (see Figure 8.1). What is the probability that the cylinder will buckle?

The critical compressive load, P_{crit}, which causes the cylinder to buckle is given by the following expression (derived from table 35 of [1]):

$$P_{crit} = 0.4 \frac{2\pi}{\sqrt{3(1-v^2)}} Et^2, \frac{r}{t} > 10, L \gg 1.72\sqrt{rt}$$

E and v are Young's modulus and Poisson's ratio, respectively, of the material from which the cylinder is made.

The condition for buckling to occur is that: $P/P_{crit} \geq 1$. There is some variability in the values of E and t, which gives rise to variability in the strengths of individual cylinders. In addition, there is some uncertainty in the value of the applied load, P. It is possible, therefore, that some cylinders will buckle and others will not. By performing a Monte-Carlo simulation, in which we select random values of each of the variable parameters, we can count the number of trials in which a cylinder buckles. This will give us the fraction of failures, or, in other words, the probability of buckling.

8.2 SIMULATION

We'll assume that the Young's modulus has a nominal value of 70 GPa with a uniform distribution having a range of ±10 GPa about this, that the wall thickness is nominally 2 mm with a Normal distribution of standard deviation 0.1 mm, and that the applied load also has a Normal distribution with a nominal value of 300 kN and a standard deviation of 10 kN. Poisson's ratio we'll set to the constant value of 0.33.

Listing 8.1 shows the program used to perform Monte-Carlo calculations on 10^6 cylinders. The in-built Python function, **randn**, generates random numbers drawn from the Normal distribution with mean zero and standard deviation unity. Because the failure probability is very small (typically, around 0.004), we need to test many more than 1000 cylinders to get a reasonable estimate, which is why we use 10^6 here.

DOI: 10.1201/9781003295235-8

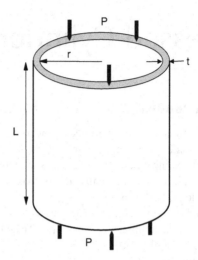

FIGURE 8.1   Thin-walled, hollow cylinder.

Listing 8.1 Buckled cylinder failure probability

```
# BuckledCylinder.py
# Calculates the probability that a hollow cylinder will
buckle
# when subjected to a compressive axial load.

import numpy as np
from numpy.random import rand, randn

# Basic data
v      =   0.33               # Poisson's ratio
tnom   =   0.002              # Nominal thickness [m]
Enom   =   70E9               # Nominal Young's modulus
[Pa]
Pnom   =   300000             # Nominal compressive
load [N]
st     =   1E-4               # Standard deviation of t
[m]
rE     =   10E9               # Half range of E [Pa]
sP     =   10000              # Standard deviation of P
[N]

# Derived data (Pcrit = Pcon.E.t^2)
Pcon = 0.4*2*np.pi/np.sqrt(3*(1-v**2))

# Number of cylinders
N = 10**6

# Random values of E (uniform), t (normal) and P (normal)
```

```
E = Enom + (2*rand(N)-1)*rE
t = tnom + randn(N)*st
P = Pnom + randn(N)*sP

# P/Pcrit
Pratio = P/(Pcon*E*t**2)

# failures = True if criterion exceeded, False if not
failures = Pratio>=1

# Overall failure probability
failprob = np.mean(failures)
```

8.3 PRECISION

By repeating the calculation shown in listing 8.1 100 times, we find the overall average buckling probability is 0.00420 with upper and lower 95% confidence limit errors on the mean of 0.00422 and 0.00419 respectively. So, we can reasonably say the overall failure probability is approximately 0.0042.

8.4 WILKS' METHOD

Our model is sufficiently small and easy to calculate that there is no difficulty in computing the probability of failure (failure fraction) for the 10^6 cylinders as specified in listing 8.1. However, suppose for a moment that this were not the case, and the model was complicated and computationally intensive; so much so that it would be impractical to do anything like a large enough number of runs to be confident of seeing any failures at all (a situation that may be true of many real-world models). Is there anything useful we could say about the failure probability based on a much smaller sample? As it happens there is: we can make use of an approach based on Order Statistics (see [2] and [3], for example).

For this we perform a much-reduced number of Monte-Carlo runs, sort the results in order of increasing propensity of failure and make a probabilistic statement about the proportion of expected failures based on the worst value observed. But how does that work? Our results are either fail or not-fail, so there will be at most two groups in which to sort them! Surely, we won't be able to say anything more than we can from the standard Monte-Carlo approach this way? That's certainly true if we only use the discrete fail/no-fail measure. However, if we were able to make use of a *continuous* measure of propensity to fail, then we could adopt an Order Statistic approach.

Fortunately, such a measure is easy to construct for this problem. We simply calculate the values of the ratio P/P_{crit}. Even if the values are all less than 1, the larger the ratio, the nearer the cylinder is to failure. So, how many Monte-Carlo trials do we run and how do we make use of the worst values we find from these trials?

According to the method described in [3], to have a confidence level, c, that the number of trials, N, required to determine the maximum value of P/P_{crit} that we find is at least as large as that which tops a fraction, f, of the true distribution of P/P_{crit}, we make use of the relationship: $c = 1 - f^N$. Rearranging this to find the number of trials we have:

$$N = \frac{\ln(1-c)}{\ln f} \tag{8.1}$$

For example, if we want to be 95% confident ($c = 0.95$) that our largest value is in the worst 5% ($f = 0.95$) of the full distribution, then the number of trials we would need to run is: $N = \ln(1 - 0.95)/\ln 0.95 \rightarrow 59$ (where we have rounded N up to the next integer value, as we can't run a fraction of a trial!).

So, if the largest value of P/P_{crit} that we find in fifty-nine trials is less than 1, then we can be 95% confident that at least 95% of our cylinders will not fail. If we were able to do another fifty-nine trials, using a different set of random numbers, we would almost certainly get a different value for the largest value of P/P_{crit}. In fact, there is a whole distribution of possible largest values of P/P_{crit}, any one of which we might get from a single set of fifty-nine trials. Figure 8.2 illustrates this distribution (dashed line) compared with the 'true' population distribution (solid line). It is clear that we are very likely to get a conservative estimate of the proportion of cylinders likely to fail from a sample of fifty-nine trials.

Figure 8.3 shows the equivalent cumulative distribution frequency curves. The dotted lines mark the value of P/P_{crit} that forms the upper boundary to 95% of the population and the lower boundary to 95% of the 'worst of 59'. Clearly then, we can be 95% confident (dashed line) that the worst value from a sample of fifty-nine trials is worse than 95% of the population (solid line) of cylinders. In the real-world situation, of course, we would only have this one worst-case value; we would know neither the population nor the sample distributions.

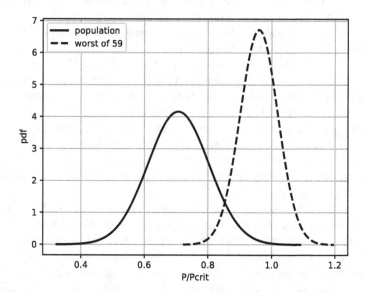

FIGURE 8.2   Population and 'worst of 59' probability density functions.

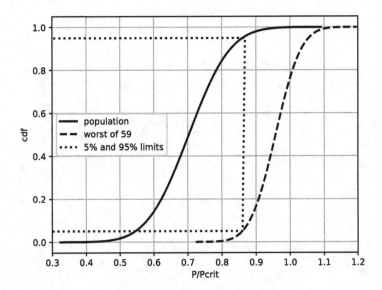

FIGURE 8.3 Population and 'worst of 59' cumulative distribution functions.

There is a potential sting in the tail with this approach, though. We can see from Figure 8.3 that there is an approximately 20% probability that our 'worst of 59' sample will indicate failure (i.e., will have a value of P/P_{crit} greater than 1). That is, we would not be able to say we are 95% confident that 95% of our cylinders will not fail. We might have to reduce our confidence level, accept that we might manufacture a lower proportion of satisfactory cylinders, or possibly even reject the design completely! (Note that the 20% probability applies to our cylinder problem here only; in general this value will be problem dependent.)

8.5 EXERCISES

8.4.1 Repeat the simulation using the method of listing 8.1, but change the wall thickness to, say, 1 mm, which should increase the probability of failure.

8.4.2 Suppose we only want to be 90% confident that our largest sample value is in the worst 5% of the full distribution. What sample size, N, would we need, and what would the 'worst of N' probability density and cumulative distribution functions look like?

REFERENCES

1. Warren C. Young (1989) ROARK'S Formulas for Stress & Strain, 6th Edition. McGraw-Hill.
2. Wilks S.S. (1942) Statistical Prediction with Special Reference to the Problem of Tolerance Limits, Annals of Mathematical Statistics, vol. 13, pp. 400–409.
3. Guba A., M. Makai, L. Pal (2003) Statistical Aspects of Best Estimate Method - I, Reliability Engineering and System Safety, vol. 80, pp. 217–232.

9 Linear Resistive Networks

9.1 CIRCUIT ANALYSIS BY RANDOM WALK

The earlier problems in this book deal with systems that are continuous, in the sense that the randomly selected trial parameters, such as radius, angle, distance travelled, etc., can take any value within a specified range. We'll now consider some systems that are discrete, such that the randomly selected trial parameters can only take one of a number of discrete values.

Picture an electrical circuit comprised of a number of interconnected electrical resistors. Focus on a particular interconnection node and the nodes and resistors to which it is directly connected. For example, let's look at a segment of the circuit where node, m, say, might be surrounded by nodes, q, r and s, to which it is connected via resistors R_1, R_2 and R_3, as in Figure 9.1.

We start conventionally by analysing this part of the circuit using Kirchhoff's current law and Ohm's law. Kirchhoff's current law says that the sum of the electrical currents out of any interconnection node must be zero (this is basically a statement of the conservation of electrical charge). So, with currents i_1, i_2 and i_3 from Figure 9.1 we have:

$$i_1 + i_2 + i_3 = 0 \tag{9.1}$$

Ohm's law relates the voltage difference across a resistor to the current flowing through it and the value of the resistance. Using V_m to represent the voltage at node m, we have for the three branches of Figure 9.1:

$$\begin{aligned}
V_m - V_q &= i_1 R_1 \\
V_m - V_r &= i_2 R_2 \\
V_m - V_s &= I_3 R_3
\end{aligned} \tag{9.2}$$

Divide each of the equations in (9.2) by its value of resistance, sum the resulting equations and make use of Equation 9.1 to obtain, after a little rearrangement:

$$V_m = \frac{\dfrac{1}{R_1} V_q + \dfrac{1}{R_2} V_r + \dfrac{1}{R_3} V_s}{\dfrac{1}{R_1} + \dfrac{1}{R_2} + \dfrac{1}{R_3}} \tag{9.3}$$

So far, we've done nothing out of the ordinary, and if we were to continue in the normal manner, we would write similar equations for all the nodes in the network, assemble them in a matrix equation and proceed to find the unknown voltages using a standard matrix solution method. However, with Monte-Carlo simulation in mind,

DOI: 10.1201/9781003295235-9

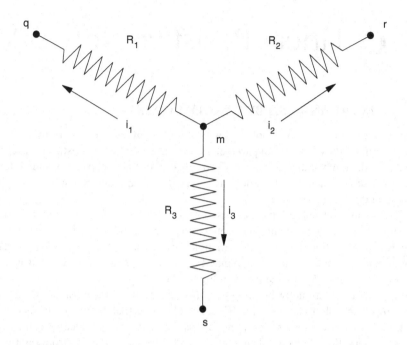

FIGURE 9.1   Segment of electrical network.

we are going to rewrite Equation 9.3 and interpret the result in a somewhat unusual way. First, we express Equation 9.3 in the form:

$$V_m = p_1 V_q + p_2 V_r + p_3 V_s \qquad (9.4)$$

where we have:

$$p_x = \frac{\dfrac{1}{R_x}}{\dfrac{1}{R_1} + \dfrac{1}{R_2} + \dfrac{1}{R_3}} \qquad (9.5)$$

for each $x = 1, 2, 3$.

Equation 9.4 says that V_m is a weighted sum of the voltages of the nodes to which it is connected. Because of the way they are defined, the weighting factors, p_x, sum to unity. That is: $p_1 + p_2 + p_3 = 1$. In this respect, they behave like probabilities, for which the sum over a complete set must equal unity. This leads us to our unusual interpretation of Equation 9.4.

Imagine that we have a virtual particle at node, m, which moves to one or other of nodes q, r and s. The choice of which node we move the particle to depends on a uniform random number in the range 0 to 1. If the value of this number is less than p_1 the particle moves to node q; if the number is between p_1 and $p_1 + p_2$ the particle moves to node r; and if the number is greater than $p_1 + p_2$ the particle moves to node s.

We record the value of voltage at the new node that the particle is now on (assuming for the moment that we know what it is). This is the result of a single step of a 'random walk'. If we repeat this for a large enough number of random walks, we should have recorded voltage V_q in a fraction p_1 of the trials, V_r in a fraction p_2 of the trials and V_s in a fraction p_3 of the trials. Hence, it follows from Equation 9.4 that the average voltage recorded will be, at least approximately, V_m.

What we have just described, of course, is a Monte-Carlo process. Let's perform a simulation for this rather trivial case of an electrical network with a single node of unknown voltage. In passing you should note that the virtual particles involved are not to be associated in any way with physical electrical particles, such as electrons; they are entirely fictional!

9.2 SINGLE-NODE SIMULATION

Figure 9.2 shows a simple electrical resistive circuit for which we know the values of all the resistors and the voltages at nodes 1, 2 and 3, and wish to find the voltage at node 4. At nodes 1, 2 and 3 we have $V_1 = 20$ volts, $V_2 = 10$ volts and $V_3 = 0$ volts. The three resistances are $R_1 = 200\ \Omega$, $R_2 = 100\ \Omega$ and $R_3 = 50\ \Omega$. Listing 9.1 details the Python program that calculates the voltage at node 4 using the Monte-Carlo simulation process described in section 9.1 above.

The average value of V_4, estimated from 1000 trials of the program of listing 9.1 is 5.7047. Our usual calculation of the standard error of this mean puts the 95% upper and lower limits as 5.7190 and 5.6904 respectively, so our best estimate would be 5.7. This compares with the true value of 5.714.

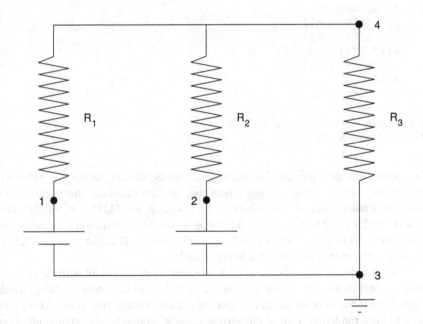

FIGURE 9.2 Simple electrical resistance circuit.

Listing 9.1 Single-node simulation

```
# SingleNode.py
# Monte-Carlo calculation of single node voltage.

import numpy as np
from numpy.random import rand

# Specify resistances, R [ohms], and known boundary node
voltages, Vb [volts]
R = np.array([200, 100, 50])
Vb = np.array([20, 10, 0])
# Inverse resistances.
InvR = 1/R
S    = np.sum(InvR)
# Weights
p    = InvR/S
pcum = np.cumsum(p)
# Number of random walks.
N = 1000
# Space for voltage estimates.
V = np.zeros(N)

# N uniform random numbers between 0 and 1
k = rand(N)

# Random walks
for t in range(N):
    if k[t]<pcum[0]:
        V[t] = Vb[0]
    elif k[t]<pcum[1]:
        V[t] = Vb[1]
    else:
        V[t] = Vb[2]

# Voltage estimate for node 4
V4 = np.sum(V)/N
```

Of course, this is a monstrously inefficient method of calculating the unknown voltage for such a simple circuit! Given that we can calculate the values of p_x using the known resistances in Equation 9.5 (to get $p_1 = 0.142857$, $p_2 = 0.2857143$, $p_3 = 0.5714286$) and that we know the voltages, we can determine V_4 directly from Equation 9.4 as $V_4 = 0.142857 \times 20 + 0.2857143 \times 10 + 0.5714286 \times 0$, or $V_4 = 5.714$. So, why bother with the Monte-Carlo approach?

Well, in general, we will have a much more complicated circuit with many more nodes for which we don't know the voltage, in which case a Monte-Carlo approach might sometimes have an advantage. For these more complicated networks we will need to keep randomly moving our virtual particle around (using random numbers suitably weighted in proportion to the inverse resistances at each of the possible moves) until eventually it reaches a node of known voltage, which we record. Repeating this for a large number of virtual particles, each of which starts from the same node, we can again estimate the voltage of that node by averaging all the recorded voltages. We won't justify that statement here; for that, look at [1] or [2], for example.

9.3 UNIT-RESISTANCE CUBE SIMULATION

Let's investigate a slightly more complicated network; that of a cube of resistors, each of 1Ω resistance, with a 1 volt potential difference applied across a pair of diagonally opposite corners. Figure 9.3 shows the arrangement, where node 7 is grounded at 0 volts and node 8 is fixed at 1 volt. What are the voltages at the other nodes?

If we imagine a virtual particle at node 1, say, we can see that although a move to node 7 will result in a known voltage, moves to nodes 2 or 6 will not. If our particle moves to node 2, say, it must then choose another node to move to (which could include moving back to node 1). It must keep repeating this process until it lands on one of the boundary nodes, 7 or 8. Having started enough trial particles from node 1 to allow us to estimate its voltage by averaging all the recorded boundary voltages reached, we then need to repeat the whole process starting from each of the other nodes for which we want to find the voltage.

Actually, the symmetry of the cube allows us to focus on only two nodes, say 1 and 2. The voltages at nodes 3 and 5 are clearly identical to that at node 1, while the voltages at nodes 4 and 6 are identical to that at node 2.

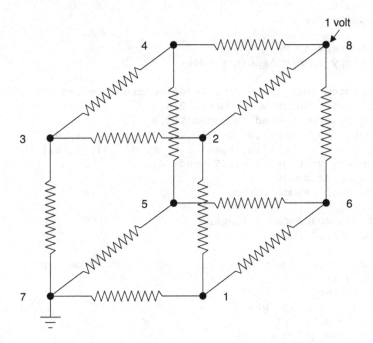

FIGURE 9.3 Cube of unit resistances.

A further simplification follows from the fact that all the resistances have the same value. This makes the probability that our virtual particle will move to any one of its three neighbouring nodes as 1/3 for each (this follows from Equation 9.5 by making all the resistances the same).

By the way, the fact that both our examples so far show exactly three branches connected to the nodes whose voltages we are trying to calculate, is coincidental and not a fundamental limitation. Clearly, the process generalises in a straightforward manner no matter how many connections a node has.

Listing 9.2 shows a Python program that can be used to find the voltages at nodes 1 and 2. When a virtual particle lands on a node it needs to know to which nodes it can now move. This information is contained in variable, **connect**. This is a 6 × 3 matrix where the ith row represents the ith node (only the first six non-boundary nodes are needed), and the three columns contain the node numbers of those nodes to which it is connected. The first two nodes are numbered 1 and 2; however, Python indices in vectors and matrices start at 0, not 1, which is why the node number is set to i+1, not just i.

Listing 9.2 Unit resistance cube simulation

```
# UnitCube.py
# Uses Random walk method to calculate voltages at nodes of a
cube of
# unit resistors subject to a voltage drop of 1 volt across
diagonally
# opposite corners.

import numpy as np
from numpy.random import randint

# Connected node list for internal nodes (i.e., nodes 1 to 6)
# node 1 is connected to nodes 2,6,7
# node 2 is connected to nodes 1,3,8  etc.
connect = np.array([[2, 6, 7], [1, 3, 8], [2, 4, 7],
                    [3, 5, 8], [4, 6, 7], [1, 5, 8]])
# Boundary voltages [node 7, node 8]
Vbnd = np.array([0, 1])
# Number of random walks
N = 1000
# Storage space for voltages
V = np.zeros([2,N])

# Random walks (only start from nodes 1 and 2 because of
symmetry )
for i in range(2):
    for t in range(N):
        node = i+1
        keepgoing = True
        while keepgoing:
```

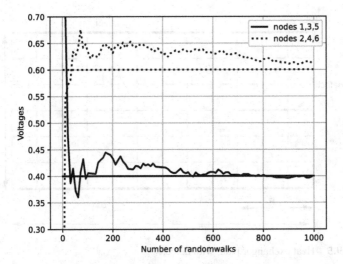

FIGURE 9.4   Unit-resistance cube voltages.

```
        p = randint(1, 4)-1
        node = connect[node-1,p]
        if node>6:
            V[i,t] = Vbnd[node-7]
            keepgoing = False
            # end of if node>6
        # end of while keepgoing
    # end of for t in range(N)
# end of for i in range(2)

# Estimate nodal voltages
V1 = np.mean(V[0,:])
V2 = np.mean(V[1,:])
```

To see how the voltages tend to converge towards a limiting value, while still fluctuating, as the number of random walks increases, Figure 9.4 shows cumulative estimates of the voltages of nodes 1 (hence 3 and 5 also) and 2 (hence 4 and 6 also). After 1000 random walks, this run estimated that $V_1 = 0.402$ volts and $V_2 = 0.613$ volts. The horizontal lines mark the true values of 0.4 volts (nodes 1, 3 and 5) and 0.6 volts (nodes 2, 4 and 6). For this example, I'll leave the estimate of the 95% confidence limits to you.

9.4   HEAT CONDUCTION SIMULATION

This approach of having virtual particles perform random walks around a network is not confined to electric circuits. Any network for which we can ascribe fixed probabilities to the paths from node to node, and for which we know values of the appropriate analogue of electrical voltage at the boundary nodes, can be treated this way.

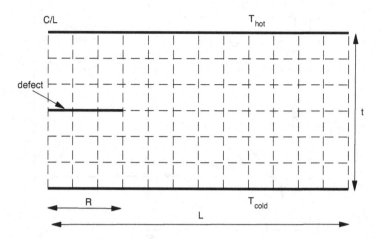

FIGURE 9.5   Heat exchanger plate with defect.

For example, imagine a metal plate in a heat exchanger, separating the hot and cold streams of fluid. Assume the faces are fixed at known hot and cold temperatures and that the plate is essentially rectangular, so that, with perfectly uniform properties there is a linear drop in temperature through the thickness of the metal. Now suppose there is a small defect in the middle of the plate: a very high thermal resistance inclusion, as illustrated in Figure 9.5. How will the temperature vary along the centre-line that passes from the hot face through the defect to the cold face? Strictly, this is a continuous problem, described by a partial differential equation. However, in practice we would probably cover the region of interest with a grid of discrete points (nodes) and solve for the temperatures at each point.

The centre-line (C/L) is that through the centre of the defect, and is the line along which we want to know the temperatures. The thickness of the metal is t units (mm, say) and the half-width of the defect is R units. To make life simple, we'll impose a square grid of side 1 unit, make the defect 10 units wide (i.e., $R = 5$), and set the thickness of the metal to be 18 units (i.e., $t = 18$). A distance, L units from the centre-line the metal is far enough away from the defect for the temperature to be unaffected by distortions introduced by its presence, so the through-thickness profile is linear again (we'll use $L = 40$). Assume the length of the defect in the 'out-of-page' direction is very much larger than R, so that we can restrict our attention to two-dimensional heat flow only. We'll set temperature T_{hot} as 100°C and T_{cold} as 20°C.

We use Fourier's law of heat conduction to relate the heat flow-rate (the thermal analogue of electrical current) between any two grid nodes to the temperature difference between them (the thermal analogue of electrical potential difference):

$$Q_{m \to N} = k_{m \to N} A_{m \to N} \left(\frac{T_m - T_N}{\delta} \right) \tag{9.6}$$

Here, $Q_{m \to N}$ is the heat flow-rate between node m and node N (we'll use N, W, S and E to indicate directions to the North, West, South and East, respectively, of a given grid node, m), $k_{m \to N}$ is the thermal conductivity, $A_{m \to N}$ is the heat transfer area and δ is the distance between the two nodes. We write analogous equations to 9.6 for directions W, S and E, sum them all, equate the result to zero and rearrange to get:

$$T_m = p_N T_N + p_W T_W + p_S T_S + p_E T_E \tag{9.7}$$

where:

$$p_X = \frac{k_{m \to X} A_{m \to X}}{k_{m \to N} A_{m \to N} + k_{m \to W} A_{m \to W} + k_{m \to S} A_{m \to S} + k_{m \to E} A_{m \to E}} \tag{9.8}$$

and X is N, W, S, E as necessary. Equation 9.7 is the analogue of equation 9.4, and 9.8 of 9.5.

At most internal nodes, those away from the defect, the thermal conductivities are the same in all directions for this problem, as are the heat transfer areas; so, for these 'standard' nodes all the probabilities reduce to $p_X = 1/4$. We'll assume the defect has an infinite thermal resistance, so any particle that lands on it immediately returns to the node from which it arrived.

Listing 9.3 shows a Python program that estimates the centre-line temperatures by following virtual particles on their random walk around the grid. Each call to the function, **randint** randomly generates one member of the set [0, 1, 2, 3]. These represent directions [North, West, South, East] respectively.

Listing 9.3 Heat conduction

```
# PlateTemperatures
# Calculation of centre-line temperatures through a plate
containing
# a defect using the random walk of virtual particles.

import numpy as np
from  numpy.random import randint

# thickness, defect half-width, plate width
t, R, L = 18, 5, 40
# hot, cold surface temperatures
Thot, Tcold = 100, 20
#  defect row
defectrow = int(t/2)
# storage for linearly falling temperatures
Tlinear = np.zeros(t+1)
# linear position temperatures
for i in range(t+1):
    Tlinear[i] = Thot - (Thot-Tcold)*i/t

# Storage for C/L temperatures
CLTemp = np.zeros(t+1)
# Fix surface temperatures
CLTemp[0], CLTemp[t] = Thot, Tcold
# Number of virtual particles
```

```
N = 10**4

# Calculate centre-line temperatures
for row in range(1,t):
    T = 0
    for trial in range(N):
        r, c = row, 0
        if r == defectrow and c == 0:
            keepgoing = False
        else: keepgoing = True
        while keepgoing:
            rc,  cc = r, c
            p = randint(4)
            # make move
            if p==0:
                r = rc-1                      # move N
                if r == defectrow and cc<R:
                    r = rc
            elif p==1:
                c = cc-1                      # move W
                if r == defectrow and c==R:
                    c = cc
            elif p==2:
                r = rc+1                      # move S
                if r == defectrow and cc<R:
                    r = rc
            else:
                c = cc+1                      # move E
            # end of if p==0
            # check boundaries
            if r<1:
                T = Thot + T
                keepgoing = False
            elif r>t:
                T = Tcold + T
                keepgoing = False
            elif c>L-1:
                T = Tlinear[r] + T
                keepgoing = False
            if c<0: c = c+1
            # end of if r<1
        # end of while keepgoing.  This particle's walk is
finished
    # end of for trial in range(N). Completed all particles
for this row
    CLTemp[row] = T/N          # average temperature for this row
# end of for row in range(1,t).  All rows completed

# set defect row temperature
CLTemp[defectrow] =
(CLTemp[defectrow-1]+CLTemp[defectrow+1])/2
```

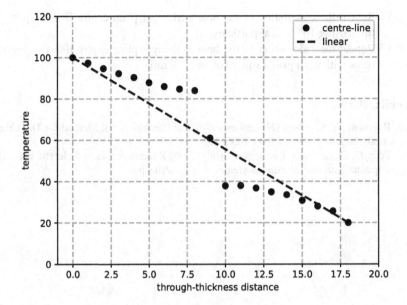

FIGURE 9.6   Centre-line temperature profile.

The resulting centre-line temperature profile is compared with the linear profile in Figure 9.6. We can see the large temperature drop across the centre of the defect. We should, of course, check the repeatability of the temperatures shown in Figure 9.6, and, if this were a real-life problem, would want to investigate the sensitivity of the result to such items as number of virtual particles, grid spacing and distance to the boundary at which we've decreed there is a linear temperature profile. Any errors are likely to be larger the closer we are to the defect, and Figure 9.6 certainly suggests there is room for improvement there, as the calculated central temperature isn't quite the value of 60°C we would expect from the symmetry of the system (for the run shown it's approximately 1°C too high). We won't pursue that any further here though.

We could extend the use of these virtual particles to calculate the temperatures at all the other nodes of the metal plate. However, there is little to be gained by doing so. Conventional solution methods, which tend to require that the parameter of inter-est be determined over the whole domain, are generally better for this situation. Any advantage to be obtained from use of the Monte-Carlo, random walk approach is likely to be achieved only where the parameter is required over a limited region of the domain (such as temperatures along the centre-line only).

9.5   EXERCISES

9.5.1 In the unit-resistance cube electrical network of section 9.3, set the resis-tance of the resistor between nodes 1 and 7 to be 2 Ω. The cube is no lon-ger symmetrical. Use the random walk method to calculate the voltages at all the internal nodes.

9.5.2 Repeat the calculation of the heat exchanger plate centre-line temperatures using 10^6 virtual particles.

9.5.3 Repeat the calculation of the heat exchanger plate centre-line temperatures with each grid a square of size 1/2 unit.

REFERENCES

1. Raymond A. Sorensen (November 1990) The Random Walk Method for DC Circuit Analysis. Am. J. Phys. 58(11), 1056–1058.
2. Peter G. Doyle and J. Laurie Snell (July 2006) Random walks and electric networks, http://math.dartmouth.edu/~doyle/docs/walks/walks.pdf

10 Magnetic Phase Transitions

10.1 ISING SPIN MODEL

We can use Monte-Carlo simulation to investigate cooperative phenomena, such as phase changes, that occur when short-range atomic or molecular interactions give rise to long-range order. For example, some magnetic materials exhibit a paramagnetic phase at high temperatures, where the magnetic structure is disordered, but undergo a phase transition to a ferromagnetic or antiferromagnetic phase at a lower temperature, where the magnetic structure is highly ordered. The transition occurs at a specific temperature, known as the Curie temperature for ferromagnets and the Néel temperature for antiferromagnets.

We'll look at a simple model in two spatial dimensions, in which we have interacting spins sited on a square lattice (it is tempting to picture these spins as atomic size bar magnets, but their main interaction is through the exchange of electrons rather than the standard dipole-dipole magnetic field, which is too weak to be of significance here). In the Ising model, each spin can take just one of two possible orientations: 'up' or 'down'. In the low temperature phase, the interaction between spins is sufficiently strong for most of the spins to align themselves in the same direction (in the case of a ferromagnet) or for adjacent spins to align themselves in opposite directions (in the case of an antiferromagnet). In the paramagnetic region the high thermal energy overcomes the tendency of the spins to align and adjacent spins are randomly arranged (see Figure 10.1). The orientation of the spins may also be affected by the application of an external magnetic field.

In all phases, there is competition between the thermal energy, which tries to disorder the arrangement of spins and the 'exchange' energy, as it is called (because it arises from adjacent atoms exchanging electrons), which tries to produce an ordered arrangement of spins. An external magnetic field also tries to impose order. We call a specific arrangement of spins a state of the system. Generally, in all phases there will be fluctuations among different states and the energies of these states may well be different. The extent of these fluctuations can give us important information about the system, as we shall see.

The overall energy of a state of a system of such interacting spins is given by [1]:

$$E = -\frac{1}{2}\sum_{i,j} J_{i,j} S_i S_j - g\mu_B H \sum_i S_i \qquad (10.1)$$

Here, S_i is the value of spin on lattice site i, and has a value +1 or −1. $J_{i,j}$ is the 'exchange' energy between sites i and j; g and μ_B are known constants (the Landé g-factor and the Bohr magneton, respectively), and H is the external magnetic field. The first sum is over all pairs of sites (the factor of 1/2 in front takes account of

DOI: 10.1201/9781003295235-10

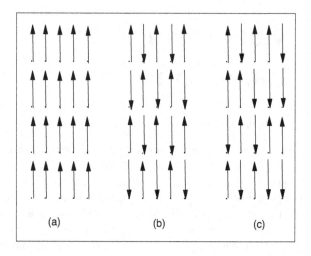

FIGURE 10.1   (a) Ferromagnet, (b) antiferromagnet, (c) paramagnet.

the fact that each pair is otherwise included twice), the second, over all sites. (Actually, this empirically based model is not a very good representation of magnetism in metals, but it represents magnetic behaviour in many non-metallic materials reasonably well.)

For the simple Ising spin model, it is normal to consider exchange interactions between nearest neighbour spins only; and we'll look at the situation in the absence of an external applied magnetic field (though, somewhat surprisingly perhaps, we will still be able to calculate the system's magnetic susceptibility), so Equation 10.1 simplifies to:

$$E = -\frac{1}{2} J \sum_{i,j} S_i S_j \qquad (10.2)$$

Now the summation is over all pairs of nearest neighbour sites and each pair has the same exchange energy, J.

10.2 PHYSICAL PROPERTIES

At a given temperature, T, the probability that the system is in a given state is proportional to the Boltzmann factor, $e^{-E/kT}$, where E is the energy of the state and k is Boltzmann's constant (see [2] for example). The constant of proportionality is obtained by realising that the system must certainly be in some state; so, by summing the Boltzmann factors over all states we obtain a normalising constant, Z. The probability, p, of being in a particular state is then:

$$p = \frac{e^{-\frac{E}{kT}}}{Z} \qquad (10.3)$$

where:

$$Z = \sum e^{-\frac{E}{kT}}$$

(10.4)

and the sum in Equation 10.4 is taken over all states. Z is known as the partition function.

The average value of some property, X, of the system is obtained from the weighted sum of the values of X in each state (we will be interested in X as energy and X as magnetisation). The weighting factors are the probabilities given in Equation 10.3.

$$\langle X \rangle = \frac{1}{Z} \sum X e^{-\frac{E}{kT}}$$

(10.5)

The use of angled brackets, $\langle X \rangle$, denotes the average value of X.

We will calculate the equilibrium values of the average energy of the system, the average magnetisation (the magnetisation, M, is simply the sum of all the spins; that is: $M = \Sigma_i S_i$), the heat capacity and the magnetic susceptibility. We'll do this for a range of temperatures from the high temperature, disordered phase down to the low temperature, ordered phase.

The heat capacity, C_v, of the system is defined by: $C_v = \partial \langle E \rangle / \partial T$. At first sight, it looks as though we might have to calculate the average energy at two slightly different temperatures to determine the heat capacity. However, we can actually find it by making use of the fact that, even at a single temperature, the system is continually fluctuating among different states. The derivation, which makes use of Equations 10.4 and 10.5, is as follows. Given that $E = (\Sigma E e^{-\frac{E}{kT}})/Z$, we have:

$$C_v = \frac{1}{Z} \frac{\partial}{\partial T} \left(\Sigma E e^{-\frac{E}{kT}} \right) - \Sigma E e^{-\frac{E}{kT}} \frac{1}{Z^2} \frac{\partial Z}{\partial T}$$

$$= \frac{1}{ZkT^2} \Sigma E^2 e^{-\frac{E}{kT}} - \Sigma E e^{-\frac{E}{kT}} \frac{1}{Z^2} \Sigma E e^{-\frac{E}{kT}} \frac{1}{kT^2}$$

(10.6)

$$= \frac{\overline{E^2} - \overline{E}^2}{kT^2}$$

The numerator of Equation 10.6 is simply the variance of E.

The magnetic susceptibility, χ, of the system is defined by: $\chi = \partial \langle M \rangle / \partial H$. Again, we can make use of the fluctuations among states to calculate the susceptibility (to derive the following result the magnetic field term must be reintroduced into the expression for energy, but the applied field can subsequently be set to zero to get the zero-field susceptibility):

$$\chi = g\mu_B \frac{\langle M^2 \rangle - \langle M \rangle^2}{kT}$$

(10.7)

The constants, k, g and μ_B, are, in effect, real-world scaling factors here. Their true values are of no great interest in our simulation below, so we'll set them all to unity. We'll also set the magnitude of J to unity, but will retain it explicitly in the simulation as +1 or –1 to allow an easy switch between ferromagnetic ($J = +1$) and antiferromagnetic ($J = -1$) systems.

Why is Monte-Carlo simulation useful here? Couldn't we calculate the averages we need directly by making use of Equation 10.5? Yes, we could do this in principle, but in practice there are likely to be far too many states to evaluate. With N spin sites, each of which can take one of two values, there are 2^N states in total. With as few as 100 sites, 2^{100} states are already too many to contemplate, and we will use 10,000 sites below! So, we'll use Monte-Carlo simulation to estimate the required averages from a, possibly large, but manageable number of samples.

10.3   THE GIBBS AND METROPOLIS ALGORITHMS

Before attempting to calculate the energy and magnetisation averages, we need to ensure the system is at equilibrium. We do this by starting from a random state, which might or might not be near equilibrium, and choosing to set the spin orientations using a weighted random choice, so as to drive towards equilibrium.

Since the probability of a spin being in one orientation is proportional to the Boltzmann factor, $e^{-\frac{E}{kT}}$, the probability that it should be in its 'up' (+1) orientation is given by:

$$p_+ = \frac{e^{-\frac{E_+}{kT}}}{e^{-\frac{E_+}{kT}} + e^{-\frac{E_-}{kT}}} = \frac{1}{1 + e^{\frac{E_- - E_+}{kT}}} \qquad (10.8)$$

where E_+ and E_- are the energies of the states with spin, i, in its 'up' and 'down' orientation respectively.

To decide if the spin should be in the 'up' orientation we could generate a uniform random number in the range 0 to 1 and compare it with probability, p_+, calculated from Equation 10.8. If it were less than p_+, we would give it an 'up' orientation, if not we would give it a 'down' orientation. By repeating this process a large number of times for spins all over the lattice, we could ensure the system eventually reached equilibrium. This process is called the Gibbs sampling algorithm.

There is, however, a somewhat more efficient method, published by Metropolis et al in 1953 [3], and known as the Metropolis algorithm. In this we take a spin and consider flipping it from its current orientation to the opposite one. The change in energy, ΔE, of such a flip, at site i, is seen from Equation 10.2 to be:

$$\Delta E = 2JS_i \sum_j S_j \qquad (10.9)$$

where the summation is over the four nearest neighbour sites (the factor of 2 arises because the spin would change magnitude by a factor of 2 in going from +1 to –1 or vice-versa). The flipped state is accepted with a probability given by:

$$p_{flip} = 1 \ if \ \Delta E \leq 0$$
$$= e^{-\frac{\Delta E}{kT}} \ if \ \Delta E > 0$$

(10.10)

That is, we flip immediately if the flipped state is of lower energy, and flip with a probability $e^{-\frac{\Delta E}{kT}}$ if it is of higher energy. Again, by repeating this process all over the lattice a large number of times we can drive the system to equilibrium. It is this Metropolis algorithm that we will adopt here.

Once at equilibrium, we continue to repeat the process, but now record values of energy and magnetisation in order to calculate the averages we need.

10.4 SIMULATION

We construct a square lattice of $L \times L$ sites (choosing $L = 100$) that initially contains a random scattering of spins of value ±1. The boundary conditions are cyclic, so that sites on the rightmost edge of the lattice are considered to be nearest neighbours of those on the leftmost edge. Similarly, the bottom sites are neighbours of the top sites.

At a specified temperature, we sweep through all the sites, using the Metropolis algorithm, Equation 10.10, at each site to determine if its spin should be flipped. This is repeated 2000 times to allow the system to reach equilibrium. At each sweep the order in which the sites are visited is randomised to reduce the potential for any spurious pattern to be imposed on the spin states.

On the assumption that the system is at equilibrium after 2000 sweeps, we repeat the process for another 1000 sweeps, this time recording the values of overall energy and magnetisation after each sweep. The 1000 values of energy and magnetisation are then used to calculate average energy, average magnetisation, heat capacity and magnetic susceptibility. Because these are all extensive properties and would change if we changed the lattice size, they are divided by the total number of sites, to provide measures that can easily be compared with those from different lattice sizes.

Listing 10.1 shows a Python program that can be used to calculate the desired values for a single temperature. We consider the ferromagnetic case, $J = +1$, here.

Note that the random order in which each site is visited is achieved by using the in-built function, **permutation**. The resulting one-dimensional list of indices must be turned into lists of row and column numbers needed to address the two-dimensional lattice sites. This is done in the user-defined function, **ind2sub**.

Listing 10.1 Ising spins

```
# Ising.py
# Two-dimensional model of an Ising-spin ferromagnet.
# Calculates energies, magnetisations, heat capacities and
magnetic
```

```
# susceptibilities of equilibrium states.

import numpy as np
from numpy.random import rand, randint, permutation

# Basic data ------------------------------------------------------------
L = 100                    # Number of spin sites on a side.
N = L**2                   # Total number of spin sites.
J = 1                      # Exchange constant +(-)
ve,ferro(antiferro)magnet
T = 2.3                    # Temperature
eqbmstart = 2000           # Number of trials to establish
equilibrium
neq       = 1000           # Number of equilibrium trials
trials    = eqbmstart + neq  # Total number of trials

# Initialisation --------------------------------------------------------
Spins = -1+2*randint(2,size=(L,L))   # lxl matrix of spins
(+/-1)
M     = np.zeros(neq)                 # storage space for
magnetisation
E     = np.zeros(neq)                 # storage space for energy

# Function to change in dices to subscripts ----------------------------
def ind2sub(indx):
    column = np.floor(indx/L);   row = indx-column*L
    return row, column

# Metropolis algorithm function -----------------------------------------
def Metropolis(rowsub, colsub, trial):
    Energy = 0
    for counter in range(N):          # for every spin site
        r = int(rowsub[counter])      # identify row and its
neighbours
        lo = r-1
        if lo<0: lo = L-1
        hi = r+1
        if hi>L-1: hi = 0
        c = int(colsub[counter])      # identify column and its
neighbours
        lf = c-1
        if lf<0: lf = L-1
        rt = c+1
        if rt>L-1: rt = 0
        # Sum neighbouring spins
        SS = Spins[lo,c]+Spins[hi,c]+Spins[r,lf]+Spins[r,rt]
        # Calculate energy change if spin were to be flipped
        DeltaE = 2*J*SS*Spins[r,c]
        # Metropolis algorithm
        if DeltaE<0 or rand()<np.exp(-DeltaE/T):   # flip spins
            Spins[r,c] = -Spins[r,c]
```

```
        if trial>=eqbmstart:           # calculate equilibrium
energy
          Energy = Energy - 0.5*J*SS*Spins[r,c]
      # end of for counter in range(N)
      return Spins, Energy

# Perform Monte-Carlo trials --------------------------------------------
for trial in range(trials):
      indx = permutation(N)             # random permutation
of 1 to N
      rowsub, colsub = ind2sub(indx)    # convert indx to
subscripts
      # Flip spins using the Metropolis algorithm
      Spins, Energy = Metropolis(rowsub, colsub, trial)
      # record energy and magnetisation
      if trial>=eqbmstart:
          t = trial-eqbmstart
          E[t] = Energy                 # energy
          M[t] = np.sum(Spins)          # magnetisation
# end of for trial in range(trials) ------------------------------------


Eav = np.mean(E)/N                      # average equilibrium
energy per spin
Cv =  (np.var(E)/T**2)/N                # specific heat capacity
per spin
Mav = np.mean(M)/N                      # magnetisation per spin
Chi = (np.var(M)/T)/N                   # magnetic
susceptibility per spin
```

The result of running this program for a range of temperatures is shown in Figure 10.2. The values of 'temperature', T, are really values of kT, but we've arbitrarily set k to unity here. As there is no preferred 'up' or 'down' direction in the model, the magnetisation at any temperature could end up as either positive or negative. For that reason, the absolute values of magnetisation are plotted in the figure.

The exact Ising model requires an infinite number of spin sites ($L = \infty$), in which case the heat capacity and magnetic susceptibility are both infinite at the Curie temperature. However, our finite approximation clearly shows the paramagnetic to ferromagnetic phase transition at a temperature close to 2.3. In fact, the transition temperature for the two-dimensional, ferromagnetic Ising model can be determined theoretically to have a value of $2/\ln\left(1+\sqrt{2}\right) \approx 2.269$ (see [4]).

Using white and black for spin-up and spin-down sites, we can see snapshots of the model spin-field at different temperatures in Figure 10.3. Temperatures $T = 1$ and $T = 2$ show the spin-field in the ferromagnetic phase, while $T = 3$ and $T = 4$ show it in the paramagnetic phase. At $T = 1$ there is not enough thermal energy to overcome the exchange energy that wants to align the spins, while at $T = 4$ the exchange energy is too small to overcome the tendency of the thermal energy to randomise them.

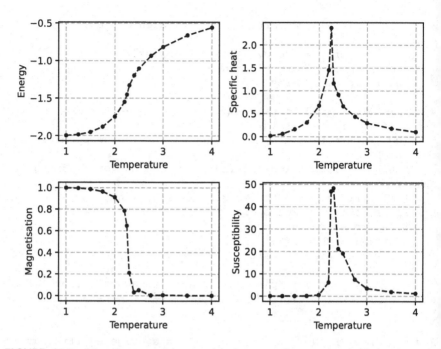

FIGURE 10.2 Ising ferromagnetic energies, heat capacities, magnetisations, magnetic susceptibilities.

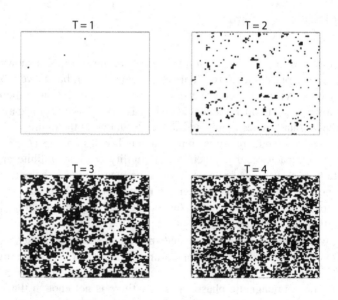

FIGURE 10.3 Snapshots of Ising ferromagnetic spin-field at four temperatures.

10.5 EXERCISES

10.5.1 Reproduce the equivalent of Figure 10.2 for an anti-ferromagnet ($J = -1$).

10.5.2 Are 2000 sweeps enough to reach equilibrium? Investigate by plotting energy and magnetisation as a function of the number of sweeps in the purported equilibrium period at a few different temperatures (ensure you choose at least one temperature near the phase transition).

10.5.3 Modify the program listed in listing 10.1 to include a magnetic field (see Equation 10.1) and set $g\mu_B H = 1$, say. What effect does this have on the energies, heat capacities, magnetisations and susceptibilities?

REFERENCES

1. Daniel C. Mattis (1965) The Theory of Magnetism. Harper & Row.
2. J.S. Dugdale (1966) Entropy and Low Temperature Physics. Hutchinson & Co. Ltd.
3. Nicholas Metropolis, Arianna W. Rosenbluth, Marshall N. Rosenbluth, Augusta H. Teller and Edward Teller (June 1953) Equation of State Calculations by Fast Computing Machines. Journal of Chemical Physics, 21 no. 6, 1087–1092.
4. Square lattice Ising model. http://en.wikipedia.org/wiki/Square_lattice_Ising_model

11 Polymer Chains

11.1 POLYMERS

Polymers are large molecules comprised of a large number of repeated units (monomers) linked end to end. There are naturally occurring polymers, such as cellulose, and artificial polymers, such as polythene (polyethylene). A simple model of a polymer assumes its linked units are each oriented completely randomly with respect to each other. This results in an essentially infinite number of possible configurations for the polymer. Therefore, it almost certainly appears as a randomly tangled mess! Figure 11.1 illustrates one possibility for a polymer constructed from 1000 units.

The dotted line in Figure 11.1 indicates the straight-line separation between the two ends of the polymer. This distance will be different, in general, for each possible configuration. We'll use Monte-Carlo simulation to calculate a distribution of values of this end-to-end distance and compare it with the following theoretical model [1]:

$$P(R) = 4\pi R^2 \left(\frac{3}{2\pi \langle r^2 \rangle} \right)^{3/2} \exp\left(-\frac{3R^2}{2\langle r^2 \rangle} \right) \tag{11.1}$$

where P is the probability density function, R is the end-to-end distance, and the mean square position of the units is $\langle r^2 \rangle = Nu^2$. N is the number of units in the chain and u is the length of each individual unit (we will assume $u = 1$ from here on).

11.2 SIMULATION

Listing 11.1 shows a Python program that calculates 10^4 samples of the end-to-end distance for a 1000 link, ideal polymer chain (we arbitrarily set the origin of the coordinate system to the start of the chain). Figure 11.2 shows a histogram of the results with the theoretical distribution superimposed. Although the end-to-end distance could, in principle, be as large as 1000 units, the chance of this happening is negligible. We can see from the figure that the polymer is likely to be tangled so much that the direct end-to-end distance will be less than 100 units, with a mean value of approximately thirty units.

Listing 11.1 Ideal polymer

```
# Polymer.py
# Calculates the average distance between the start and end of
an
# ideal polymer chain. Each link is assumed to have unit
length.
```

DOI: 10.1201/9781003295235-11

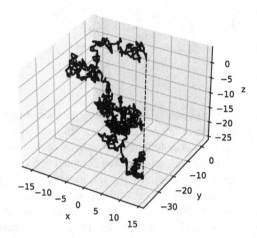

FIGURE 11.1 Simple polymer chain.

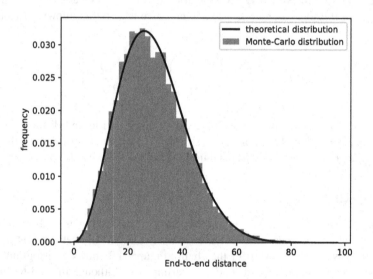

FIGURE 11.2 Polymer end-to-end distribution.

```
import numpy as np
from numpy.random import rand

 # Number of links in the chain
Nlinks = 1000
# Function to calculate distance between start and end of
chain
def chain(Nlinks):
    theta = np.pi*rand(Nlinks)
    phi = 2*np.pi*rand(Nlinks)
```

```
dx = np.sin(theta)*np.cos(phi)
dy = np.sin(theta)*np.sin(phi)
dz = np.cos(theta)
x = np.sum(dx)
y = np.sum(dy)
z = np.sum(dz)
R = np.sqrt(x**2 + y**2 + z**2)
return R

# Number of Monte-Carlo trials
M = 10000
R = np.zeros(M)
for i in range(M):
    R[i] = chain(Nlinks)
Rmean = np.mean(R)
```

For real polymers, adjacent links don't have the complete freedom of orientation allowed in the case of the ideal model. We can simulate a simple constrained orientation model by restricting the relative orientation of adjacent links. We'll prevent adjacent links from orienting themselves such that there is an acute angle smaller than, say 45° between them. We can do this by checking the angle between adjacent links after choosing a random orientation and repeatedly choosing the random orientation until that angle is larger than 45°. In effect, the links are chosen from a unit sphere with a missing cone. If we think of the links as going from the start to the end of the polymer in sequence, then the axis of each missing cone is along the line of the previous link.

Figure 11.3 shows the resulting histogram of 10^4 repeats of the end-to-end distances, with a 1000-link chain. It is plotted alongside the theoretical distribution

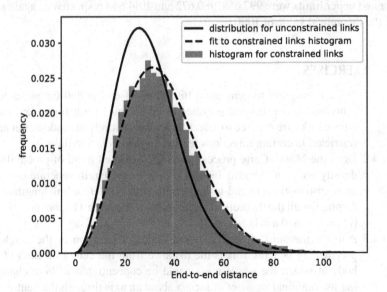

FIGURE 11.3 Polymer end-to-end distribution with constrained link orientation.

from the ideal, unconstrained distances, which shows that, as we might expect, the end-to-end distances for the constrained polymer tend to be larger than those for the ideal polymer. In fact, the mean value has increased from close to 30 to close to 35 units. The mean square value has increased from close to 1000, to about 1480 units squared. The dashed line has been calculated assuming the constrained distribution follows the theoretical probability density function of Equation 11.1, with the mean square distance of 1480 instead of 1000 (though I don't know that there is any theoretical justification for this!).

11.3 PRECISION

It is clear from Figure 11.2 that the Monte-Carlo histogram provides a good match to the theoretical distribution. However, we should confirm by comparing the value of the mean end-to-end distance with the value obtained from the theoretical distribution, and also by comparing the value of $\langle r^2 \rangle$ with the number of links (remembering that our links are of unit length).

We could do this as in previous chapters by running the polymer program many more times. An alternative, which we will adopt here, is to consider the 10^4 trials of a single run of the program as, say, 100 independent runs of 100 trials. We find the appropriate mean values from the 100 runs and determine the corresponding standard errors from the distribution of these means (assuming, as always, that they are distributed normally). Doing this for one set of 10^4 trials I obtained a value of 28.925 for the mean end-to-end distance, with a standard error of 0.122, which leads to 95% confidence limits of 28.685 to 29.165. The theoretical value, obtained from $\int_0^{1000} P(R) \times R dR$ is 29.135, so is covered by our limits.

From the same run of the program, the mean square distances and their 95% lower and upper limits were 997.658, 980.672 and 1014.644 respectively, again spanning the theoretical value of 1000.

11.4 EXERCISES

11.4.1 Write a program to generate a histogram of end-to-end distances for a non-ideal, constrained polymer of 1000 unit length links; i.e., one whose links are not free to orient themselves entirely at random, but are restricted to certain directions relative to their adjacent links.

11.4.2 Does the Monte-Carlo process show the expected trend of probability density with link length? Instead of using unit length generate one or more distributions of end-to-end length with a different link length and compare with the theoretical distribution of Equation 11.1, but now with $\langle r^2 \rangle = Nu^2$ and $u \neq 1$.

11.4.3 Another measure of interest is the radius of gyration of the tangled polymer. In general, this is the distance from the centre of mass of a body at which the whole mass could be concentrated without changing its rotational moment of inertia about an axis through the centre of mass point. For the polymer, this is the root mean square distance of the individual units from the centre of mass, given by:

$$R_{gyration} = \sqrt{\sum_{i=1}^{N}(r_i - r_{CM})^2 / N}$$

Calculate a histogram of the radii of gyration for a 1000-link ideal polymer (i.e., one with no constraints on the orientation of its links).

REFERENCE

1. Random coil. https://en.wikipedia.org/wiki/Random_coil

12 Solutions to Selected Exercises

12.1 EXERCISE 2.7.4

Picture n bins and n balls. Construct a Monte-Carlo simulation program that randomly scatters the n balls into the bins. Count the number of bins that don't contain any balls, n_0. Plot the ratio n/n_0 as a function of n. To what well known constant does the ratio seem to be trying for (no, this time it's not π!)?

Listing 12.1 contains a Python program that performs the appropriate calculations.

Listing 12.1 Balls in bins

```python
# Prob2_7_4.py
# Distribute N balls into N bins.

import numpy as np
from numpy.random import randint

# lower, upper limits, nbr of steps
lo, hi, steps = 100, 1000, 20
# nbr of balls and bins
N = np.linspace(lo, hi, steps)
# storage space
R = np.zeros(steps)

# loop through each N
for i in range(steps):
    n = int(N[i])
    bins = np.zeros(n)
    for j in range(n):
        r = randint(0,n)
        bins[r] = 1
    n0 = n - sum(bins)
    R[i] = n/n0
```

Figure 12.1 shows the values of the n/n_0 as a function of the number of bins. The ratio seems to be fluctuating about Euler's number, $e \approx 2.718$. Let's repeat the calculation of 1000 balls and bins 1000 times to get a more precise estimate. Figure 12.2 shows the resulting histogram with a superimposed normal curve. The 95% confidence limits on the mean of 2.722 are $\mu - 1.96se = 2.7176$ and $\mu + 1.96se = 2.7266$. So, the limits straddle Euler's number, though we could probably only be confident in specifying a value of 2.7 from this calculation!

DOI: 10.1201/9781003295235-12

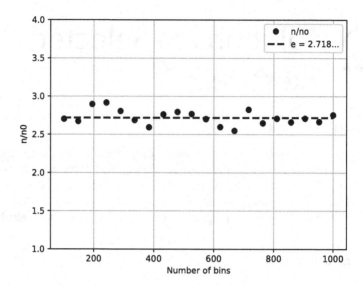

FIGURE 12.1 Ratio n/n_0 against number of bins.

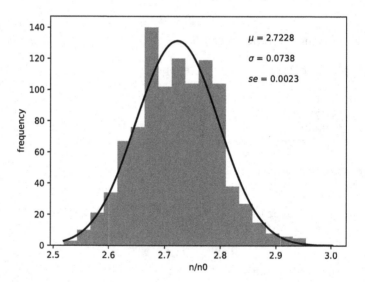

FIGURE 12.2 Histogram of n/n_0 for $n = 1000$.

12.2 EXERCISE 3.5.2

Construct a Monte-Carlo simulation program to calculate the area under the curve of the function $y = \sqrt{x}/(1 + 0.1\sin \pi x)$ between the limits $x = 0$ and $x = 1$ (the area is 0.626, correct to three significant figures).

We can do this with the simple program shown in listing 12.2. The resulting values for the mean, lower 95% limit and upper 95% limit from a run of this code were 0.6266, 0.6257 and 0.6275, respectively.

Listing 12.2 Integral of problem 3.5.2

```
# Exercise3_5_2.py
# Monte-Carlo calculation of integral of y = sqrt(x)/
(1+0.1*sin(pi*x))
# from x = 0 to x = 1

import numpy as np
from  numpy.random import rand

def y_fn(x):
    return np.sqrt(x)/(1+0.1*np.sin(np.pi*x))

trials = 500
area = np.zeros(trials)
for i in range(trials):
    N = 500
    x = rand(N)
    yave = np.mean(y_fn(x))
    area[i] = yave*1

# summary statistics
mu = np.mean(area)
sigma = np.std(area)
se = sigma/np.sqrt(trials)
lo95 = mu - 1.96*se
hi95 = mu + 1.95*se
```

12.3   EXERCISE 4.4.1

4.4.1 Calculate the view factor from horizontal to vertical when the two surfaces are infinitely wide (i.e., $Y = \infty$ in Figure 4.1). (The true value is $(1/2)\left(1 + Z/X - \sqrt{(1+(Z/X)^2)}\right)$ using the notation of Figure 4.1)

Because of the symmetry in the limit of infinite Y, this is equivalent to finding the view factor of a two-dimensional slice through the surfaces in the X-Z plane. One approach to tackling this would be to develop an entirely new program to do the calculations. However, we could also make direct use of the existing program shown in listing 4.1. All we need to do is to replace the value of Y with infinity! OK, we can't do that literally, but, perhaps, with a large enough finite value for Y we could get a result that is close enough. The problem is: how do we decide on a value that is large enough?

Well, we know that a view factor cannot be greater than 1 (a view factor of 1 means that *all* of the source radiation has been intercepted), so as Y increases the corresponding view factor must start to plateau. In principle, this plateauing might occur at such a large value of Y that it would be impractical to run the listing 4.1 program. However, it is a simple matter to try a few reasonably sized values to see if we can get acceptably close to the plateau. Figure 12.3 shows that with a Y value of a few

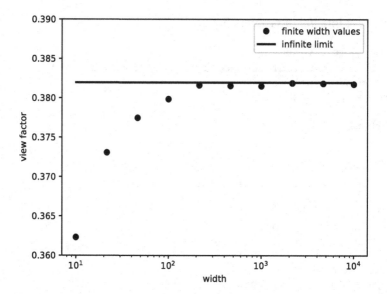

FIGURE 12.3   Perpendicular surfaces view factors as a function of lateral width.

thousand units of length we do indeed reach a plateau, with view factors which are very close to the analytical result (of course, we should really look at the variability of these values, though I won't do that here).

12.4   EXERCISE 5.4.2

How does the neutral axis offset, h, vary with the ratio R_c/R? Try a range of ratios from, say, $R_c/R = 1$ to $R_c/R = 5$.

It is a simple matter to modify the program of listing 5.1 to cover a range of values of R_c/R. Listing 12.3 shows a Python program that does this. I've included the calculation of the true values. Typical results are plotted in Figure 12.4, where the Monte-Carlo simulation results are compared with the true values.

Listing 12.3 Varying neutral axis offset

```
# CurvedBeam2.py
# Estimates distance, h, between centroidal and neutral axes
# of a curved beam of circular cross-section

import numpy as np
from  numpy.random import rand

R = 1                              # Radius of circle, R
N = 5000                           # Number of points
n = 20                             # n ratios
Rc = np.linspace(1,5,n)      # distance from centre of bending
```

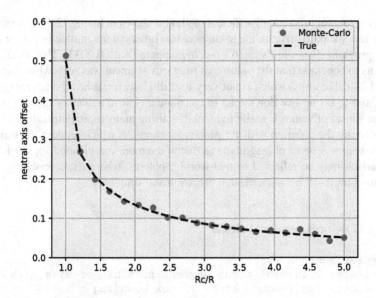

FIGURE 12.4   Neutral axis offset as a function of R_c/R.

```
h = np.zeros(n)

# function to calculate offset for a given value of Rc
def offset(Rc):
    r = np.sqrt(rand(N)*R**2)          # radii
    theta = rand(N)*2*np.pi            # angles
    s = r*np.sin(theta)                        # displacements from
centroidal axis
    den = 1/(Rc - s)                           # integrand
denominator
    num = s*den                                # integrand
numerator
    return np.trapz(num)/np.trapz(den)         # neutral axis
offset

for i in range(n):                             # loop through n
values of Rc
    h[i] = offset(Rc[i])

# True values
ht = Rc - 0.5*R**2/(Rc - np.sqrt(Rc**2 - R**2))
```

12.5 EXERCISE 6.4.2

How does the method used in listing 6.1 compare with the straightforward approach of enclosing the torus segment in a rectangular box that has sides comprising the x-y planes at z-values of 1 and –1, the x-z planes at y-values of 4 and –4 and the y-z planes at x-values of 1 and 4, within which the random points are scattered?

A program for this rather inefficient method is shown in listing 12.4. As written, it results in a standard error on the mean over two orders of magnitude larger than that of the program shown in listing 6.1 (~0.76 compared with 0.0033). This would only allow us to conclude that the volume of the torus segment was somewhere between 21 and 24 cubic units, which is not very useful! However, it has the advantage that it is easier to set up (we don't need to calculate awkward angles of the end pieces) and the spread of values is easily tightened by using many more points. It is a trivial matter to run the program with 10^8 points, for example, which reduces the standard error to the same order of magnitude as that of the more complicated approach. This comparison may be reflected in real-world problems, where there could also be a need for a trade-off between desired precision and simplicity.

Listing 12.4 Torus segment volume

```
# TorusSegmentBox.py
# Estimates volume of torus segment from hit-or-miss method
# by scattering random points in box bounding the
# whole torus segment.

import numpy as np
from  numpy.random import rand

# Radius of torus centreline and torus "tube"
Rc, Rt = 3, 1
# Volume of bounding box
Vbox = 2*3*8
# Number of points
N = 1000

# Generate N random values of x, y, z and r
x = 1 + rand(N)*3
y = -4 + rand(N)*8
z = -1 + rand(N)*2
r = np.sqrt(x**2 + y**2)   # radius

# Determine fraction of points within the torus segment
hits = np.sqrt( (r-Rc)**2 + z**2 )<=Rt
f = np.mean(hits)

# Estimate of torus segment volume
V = f*Vbox
```

12.6   EXERCISE 7.6.2

A smoke particle diffuses through a room due to isotropic, random collisions with air molecules. In spherical polar coordinates a volume element, dV, is given by $dV = r^2 \sin\theta\, d\theta d\phi dr = -r^2 d(\cos\theta) d\phi dr$, *where r is the distance travelled between*

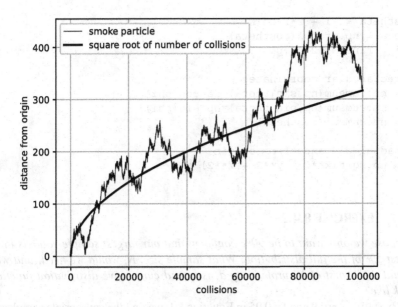

FIGURE 12.5   Random walk of smoke particle.

collisions, and θ and φ lie between 0 and π and 0 and 2π, respectively. This means $\cos\theta$ *must be chosen uniformly at random between −1 and +1, and φ between 0 and 2π. Develop a Monte-Carlo simulation program to calculate the average radial distance (in mean free paths), travelled by the smoke particle, from an arbitrary starting point, as a function of the number of collisions it experiences. For simplicity assume the distance travelled between each collision is exactly one mean free path.*

With the fixed step-size of one mean free path, this is just a three-dimensional version of the so-called 'drunkard's walk'. It is simply programmed as seen in listing 12.5. The resulting fluctuating distance from the origin is plotted in Figure 12.5, where we can see that, on average the distance increases roughly in proportion to the square root of the number of collisions, as we would expect for the 'drunkard's walk'.

Listing 12.5 Diffusion of smoke particles

```
# Exercise_7_6_2
# Diffusion of smoke particle (3D drunkard's walk)

import numpy as np
from  numpy.random import rand

# number of steps
N = 10**5

# random angles
```

```
costheta = -1 + 2*rand(N)
theta = np.arccos(costheta)
phi = rand(N)*2*np.pi

# rectangular coordinates
x = np.cumsum(np.sin(theta)*np.cos(phi))
y = np.cumsum(np.sin(theta)*np.sin(phi))
z = np.cumsum(costheta)

# radial distance from origin
r = np.sqrt(x**2 + y**2 + z**2)
```

12.7 EXERCISE 8.4.2

Suppose we only want to be 90% confident that our largest sample value is in the worst 5% of the full distribution. What sample size, N, would we need, and what would the 'worst of N' probability density and cumulative distribution functions look like?

By setting $c = 0.9$ and $f = 0.95$ in Equation 8.1, we find the appropriate sample size to be $N = \ln(1-0.9)/\ln 0.95 \to 45$ when rounded up to the nearest integer. It is then a straightforward matter to generate the probability density and cumulative distribution functions curves as for the 95% confidence case. The resulting curves are shown in Figure 12.6.

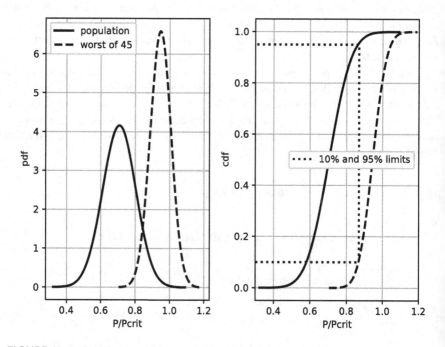

FIGURE 12.6 Population and 'worst of 45' pdfs and cdfs.

12.8   EXERCISE 9.5.1

In the unit-resistance cube electrical network of section 9.3, set the resistance of the resistor between nodes 1 and 7 to be 2 Ω. The cube is no longer symmetrical. Use the random walk method to calculate the voltages at all the internal nodes.

When our virtual particle is now on node 1, the probabilities of it moving to nodes 2, 6 and 7 are obtained from Equation 9.5 as $(1/1)/([1/2+1/1+1/1]) = 0.4$, $(1/1)/[1/2+1/1+1/1] = 0.4$ and $(1/2)/([1/2+1/1+1/1]) = 0.2$, respectively. The probabilities of moving while on the other nodes remain as they were for the unit cube situation. We therefore need to modify the program shown in listing 9.1 to reflect the changes. This is done by calculating a random integer from 0 to 4 for node 1. If the integer is 0 or 1, we move the particle to node 2; if it's 2 or 3 we move to node 6 and if it's 4 we move to node 7. Hence, each of the moves from node 1 to nodes 2 and 6 have twice the probability of a move from node 1 to node 7. Listing 12.6 shows the revised section of the program that performs the random walks.

Listing 12.6 Asymmetric cube

```
# Random walks
for i in range(6):                     # For each starting
node in turn
      for t in range(N):               # For each random walk
in turn
          node = i+1                   # keep track of
current node
          keepgoing = True
          while keepgoing:             # while not on
boundary node
              if node>1:
                  p = randint(1, 4)-1  # randomly select next
node
              else:
                  p = randint(1, 6)-1  # on node 1
                  if p<2:
                      p = 0            # go to node 2
                  elif p<4:
                      p = 1            # go to node 6
                  else:
                      p = 2            # go to node 7
              node = connect[node-1,p] # jump to next node
              if node>6:               # if on boundary node
                  V[i,t] = Vbnd[node-7]  # update voltage
                  keepgoing = False    # end this particle's
walk
              # end of if node>6
          # end of while keepgoing
      # end of for t in range(N)
# end of for i in range(6)
```

You should find that the individual node voltages are now different because the symmetry is lost. Node 1 is at a somewhat higher voltage than for the unit cube case because it is further away from 'ground'.

12.9   EXERCISE 10.5.3

Modify the program listed in listing 10.1 to include a magnetic field (see Equation 10.1) and set $g\mu_B H = 1$, say. What effect does this have on the energies, heat capacities, magnetisations and susceptibilities?

It is a straightforward matter to add a constant, say, $g\mu H = 1$, to the coding of listing 10.1, then alter the expressions, $DeltaE = 2 * J * SS * Spins[r,c]$, to $DeltaE = 2*J*SS*Spins[r,c]+2*g\mu H*Spins[r,c]$, and $Energy = Energy - 0.5*J*Spins[r,c]$, to $Energy = Energy - 0.5*J*Spins[r,c] - g\mu H*Spins[r,c]$.

The results can be seen in Figure 12.7. Not surprisingly, the applied magnetic field maintains the spin alignment against the thermal disruption to higher temperatures than when it is absent.

12.10   EXERCISE 11.4.3

Another measure of interest is the radius of gyration of the tangled polymer. In general, this is the distance from the centre of mass of a body at which the whole mass could be concentrated without changing its rotational moment of inertia about

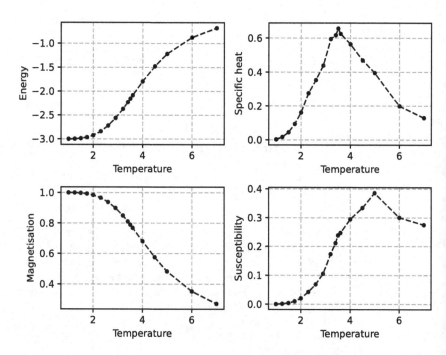

FIGURE 12.7   Ising ferromagnetic properties with applied magnetic field.

an axis through the centre of mass point. For the polymer, this is the root mean square distance of the individual units from the centre of mass, given by:

$$R_{gyration} = \sqrt{\sum_{i=1}^{N}(r_i - r_{CM})^2 / N}$$

Calculate a histogram of the radii of gyration for a 1000 link ideal polymer (i.e., one with no constraints on the orientation of its links).

This can be done by making small modifications to the program shown in listing 11.1. The principal difference is the use of the cumulative sum (**np.cumsum**) instead of the total sum (**np.sum**). Listing 12.7 shows the modified program.

Listing 12.7 Polymer radius of gyration

```
# Polymer2.py
# Calculates the radius of gyration of an ideal polymer chain.
# Each link is assumed to have unit length.

import numpy as np
from numpy.random import rand

 # Number of links in the chain
Nlinks = 1000

# Function to calculate radius of gyration
def chain(Nlinks):
    theta = np.pi*rand(Nlinks)
    phi = 2*np.pi*rand(Nlinks)
    dx = np.sin(theta)*np.cos(phi)
    dy = np.sin(theta)*np.sin(phi)
    dz = np.cos(theta)
    x = np.cumsum(dx)
    y = np.cumsum(dy)
    z = np.cumsum(dz)
    Rcmx = np.mean(x)
    Rcmy = np.mean(y)
    Rcmz = np.mean(z)
    Rcm = np.sqrt(Rcmx**2 + Rcmy**2 + Rcmz**2)
    R = np.sqrt(x**2 + y**2 + z**2)
    Rgyration = np.sqrt(np.sum((R-Rcm)**2)/Nlinks)
    return Rgyration

# Number of Monte-Carlo trials
M = 10000
Rgyration = np.zeros(M)

for i in range(M):
    Rgyration[i] = chain(Nlinks)
```

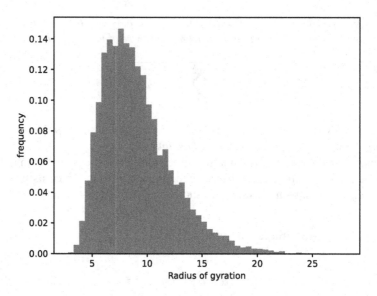

FIGURE 12.8   Polymer radius of gyration histogram.

Figure 12.8 shows the resulting histogram. As would be expected, the radius of gyration tends to be smaller than the end-to-end distance.

Appendix A: Random Numbers

A.1 INTRODUCTION

Monte-Carlo methods rely on the ability to produce streams of random numbers quickly. Although there are physical processes, such as spinning wheels or electrical noise, that can be used to generate these numbers, in practice computer programs use software algorithms to do so. As these algorithms are deterministic, the 'random' numbers they produce are, in fact, 'pseudo-random'. This means that, starting from the same initial 'seed' number, the same stream of random numbers will always be produced. Strangely, this can be an advantage when constructing a computer simulation, as it helps in debugging the program – different outputs can be identified as resulting from coding changes rather than from the use of a different stream of random numbers.

Most pseudo-random number generators rely initially on the generation of uniformly distributed numbers on the interval 0–1, regardless of the type of distribution ultimately required, so we will be briefly describe a common method for generating these. Normally distributed random numbers are also frequently required, so we will also describe a method for generating these.

Modern software programs generally make use of more complicated pseudo-random number generators than those described below. They also tend to have built-in routines for generating random numbers from distributions other than just uniform and normal (for example, Python has routines for uniform, normal, lognormal, triangular, Weibull and several others).

A.2 UNIFORMLY DISTRIBUTED RANDOM NUMBERS

A common method for generating uniform random number sequences is the linear congruential method. This generates a sequence of positive integers, x_i, $i = 1,2,...$, from an initial 'seed' value, x_0, using the recurrence relationship:

$$x_i = (a \times x_{i-1} + c) \bmod m \tag{A.1}$$

where the multiplier, a, the increment, c, and the modulus, m, all integers, need to be chosen carefully if the resulting number sequence is to have good random properties.

The values of x_0, a and c lie in the interval 0 to $m - 1$, and the pseudo-random numbers are given by the ratios x_i/m. As long as m is large the difference between successive values of $1/m$, $2/m$, $3/m$, ... etc., is very small so that the ratios can be treated as though they are continuously distributed.

DOI: 10.1201/9781003295235-13

For any such sequence, there is an interval after which the numbers start to repeat. We would like this interval – its period – to be very large; ideally, equal to m. For this to happen, the following conditions must hold:

c and m are relatively prime
$a - 1$ is divisible by all prime factors of m
$a - 1$ is divisible by 4 if m is divisible by 4

An example of a good generator, for 32-bit systems, is given by the values $m = 2^{31}$, $a = 906185749$, $c = 1$.

Setting $c = 0$ in equation A.1, we get a multiplicative congruential generator. The choice of values for a and m must be different from the above to ensure a good sequence of pseudo-random numbers. The values $m = 2^{31} - 1$, $a = 16807$ are known to give rise to desirable random sequence properties.

There are many other types of uniform random number generators, such as shift register algorithms and lagged Fibonacci generators (see [1] for example).

A.3   NORMALLY DISTRIBUTED RANDOM NUMBERS

We'll describe a well-known approach to generating normally distributed random numbers known as the Box-Müller method. Note that the standard normal probability density function of a random variable with zero mean and unit standard deviation is given by:

$$f(z) = \frac{1}{\sqrt{2\pi}} e^{-z^2/2}$$

where $-\infty < z < \infty$.

The joint probability density function of two independent standard normal random variables, x and y is then:

$$f(x,y) = \frac{1}{\sqrt{2\pi}} e^{-(x^2+y^2)/2}$$

If we picture this as a surface over the whole Cartesian plane, then the probability of being at an infinitesimal region, $dxdy$ around (x,y) is $f(x,y)dxdy$. Transforming to polar coordinates (remembering that $x^2 + y^2 = r^2$ and an infinitesimal area is $rdrd\theta$) we have:

$$f(r,\theta) rdrd\theta = \frac{1}{\sqrt{2\pi}} e^{-r^2/2} rdrd\theta$$

where $0 < r < \infty$ and $0 \le \theta \le 2\pi$. Given two uniformly distributed random numbers, r_1 and r_2 we set $r^2/2 = -\ln r_1$, or $r = \sqrt{-2\ln r_1}$, and $\theta = 2\pi r_2$. Transforming back to Cartesian coordinates we have: $x = \sqrt{-2\ln r_1} \cos(2\pi r_2)$ and $y = \sqrt{-2\ln r_1} \sin(2\pi r_2)$, so this gives us two normally distributed random numbers.

There are improvements possible to this basic approach (see [2] for example), and many other ways of generating normally distributed random numbers.

REFERENCES

1. David P. Landau, Kurt Binder (2009) A Guide to Monte-Carlo Simulations in Statistical Physics. Cambridge University Press.
2. Sheldon M. Ross (2006) Simulation. Elsevier Academic Press.

Appendix B: Variance Reduction

B.1 INTRODUCTION

The naïve approach to Monte-Carlo simulation taken in most chapters of this book can sometimes be inefficient in real-world problems. That is, the resulting precision might be unacceptably large for a reasonable number of trials, or the number of trials might be unacceptably large for a desired precision. There are a number of possible variance reduction techniques that might be applicable in certain circumstances that could reduce the computational effort required to achieve an acceptable balance. We'll briefly describe three such methods here.

Monte-Carlo methods come into their own for high-dimensional problems, but to keep this chapter simple we'll restrict ourselves to one-dimensional problems only. In particular, let's assume we want to calculate the expected value of a continuous function of x, $h(x)$, where the values of x have a probability distribution function, $p(x)$. Then we can write:

$$E\big(h(x)\big) = \int h(x)\,p(x)\,dx \qquad \text{(B.1)}$$

where $E(h(x))$ is the expected value of $h(x)$.

For our Monte-Carlo approximation to the expected value of $h(x)$ we have:

$$E\big(h(x)\big) \approx \frac{1}{n}\sum_{i=1}^{n} h(x_i) \qquad \text{(B.2)}$$

where n is a large number and the values of x_i are drawn from $p(x)$.

B.2 IMPORTANCE WEIGHTING

To take a specific example let's just set $h(x) = x$, where the probability density function, $p(x)$, is given by the 'triangular' distribution illustrated in Figure B.1. We are going to estimate the expected value of x over the range $x = 0$ to $x = 4$. In general, to sample from a distribution we calculate a uniform random number, z, in the range $z = 0$ to $z = 1$, and then use this in the appropriate inverse *cumulative* distribution function. For now, we are going to assume that this is difficult to do for the triangular distribution shown in Figure B.1 (it isn't, but we are going to pretend it is!), so we'd like to generate random values of x using a distribution from which it is easy to sample, $q(x)$ (which we'll specify shortly).

We can do this by modifying our Equations B.1 and B.2 as follows:

$$E(x) = \int x\,\frac{p(x)}{q(x)}\,q(x)\,dx \qquad \text{(B.3)}$$

DOI: 10.1201/9781003295235-14

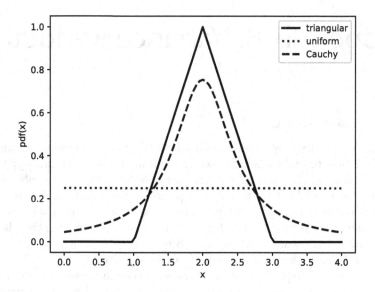

FIGURE B.1   Probability density functions.

and:

$$E(x) \approx \frac{1}{n} \sum_{i=1}^{n} x_i \frac{p(x_i)}{q(x_i)} \tag{B.4}$$

where the values of x_i are now drawn from $q(x)$, rather than $p(x)$ (and we've replaced $h(x)$ by x).

All we've done, of course, is to multiply the integrand of Equation B.1 by unity! This won't change the expected value of $h(x)$, but by expressing our Monte-Carlo approximation as in Equation B.4 we can (must) generate random values using the easier to sample from distribution, $q(x)$. The term $p(x)/q(x)$ is an importance weighting term, required to correct for the fact that we're not generating random numbers from $p(x)$. So, the question is: what should we choose for $q(x)$?

All the probability density functions, $q(x)$, that we *could* choose, will tend to home in on the true expected value as we increase the number of trials, n. However, the variance for a given number of trials can vary greatly, depending on the function chosen. To illustrate this, we'll compare two different functions for $q(x)$ here, namely, the uniform probability density function and the Cauchy probability density function, both of which are illustrated in Figure B.1 (the Cauchy pdf shown is given by $q(x) = \tan^{-1}(4) \times (1 + (2x - 4)^2)^{-1})$.

For each of the two density functions we'll estimate the expected value of x using 100 samples. We'll repeat this 100 times and plot the resulting estimates. The results are shown in Figure B.2, where we can clearly see that both functions produce expected values scattered about the true value of 2, but that the Cauchy function has a smaller variance. In general, the more we can make $q(x)$ look like $p(x)$, while still being easy to sample from, the smaller the variance is likely to be for a given value of n.

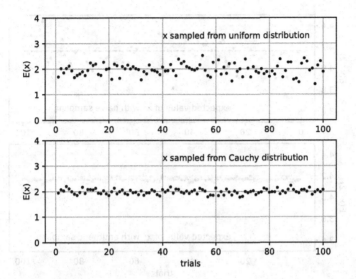

FIGURE B.2 Estimates of expected values of x.

B.3 ANTITHETIC VARIATES

If we choose two sets of random numbers, say x_i and y_i, to determine the expected value of $h(x)$, separately, then, as long as n is large enough, we should find that $E\big(h(x_i)\big) \approx E\big(h(y_i)\big)$. If the x_i and y_i are chosen independently, then the variance of the sum of the two estimates will be given by $var\big(h(x_i)+h(y_i)\big) \approx var\big(h(x_i)\big)+ var\big(h(y_i)\big)$. However, if the x_i and y_i are *not* independent, then the variance of the sum will include the covariance:

$$var\big(h(x_i)+h(y_i)\big) \approx var\big(h(x_i)\big)+var\big(h(y_i)\big)+2 \times covar\big(h(x_i)h(y)\big)$$

For situations where the covariance is negative, the overall variance of the sum will be less than the sum of the separate variances. By averaging $E\big(h(x_i)\big)$ and $E\big(h(y_i)\big)$ in such a situation, we can reduce the variance of the estimate for a given n (or reduce n for a given variance, of course). By choosing random variables appropriately, we can achieve this. Random variables defined over the same probability space are said to be antithetic if they have the same distribution and their covariance is negative. Let's see how this might work.

We'll use the same probability density function, $p(x)$, as in the previous section, but this time we'll calculate the expected value of x^2, i.e., we'll have $h(x) = x^2$. Also, we'll use the actual inverse cumulative distribution function appropriate to $p(x)$, rather than assuming it's too difficult to obtain (so that a random value of x is given by:

$$x = 1+\sqrt{2z} \qquad z < 0.5$$
$$= 3-\sqrt{2(1-z)} \qquad z \geq 0.5$$

where z is a uniform random number between 0 and 1).

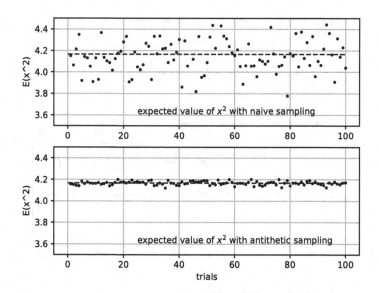

FIGURE B.3   Estimates of expected values of x^2.

For each random unit variable, z, that we use to generate values of x, and hence x^2, we'll also use $1 - z$. We'll take 100 values of z (and hence $1 - z$) and average the corresponding values of x^2. As in the previous section, we'll repeat this 100 times and plot the values for a visual indication of the scatter. Figure B.3 shows how this scatter compares with that obtained by the direct, or naive, Monte-Carlo calculation for the same number of random numbers, n. The true expected value of 4.167 is plotted as a dashed line.

B.4   CONTROL VARIATES

Suppose, once again, we have a function $h(x)$, where the x's are subject to a probability density function, $p(x)$, for which we intend to find its expectation value using Monte-Carlo simulation. Suppose, also, that there is another function, $g(x)$, where the x's are subject to the *same* probability density function $p(x)$, but for which we know its *exact* expectation value, g_{av}. Then, in general, we can express the approximate expectation value of $h(x)$ as follows:

$$E\big(h(x)\big) \approx \frac{1}{n}\sum_{i=1}^{n}h(x_i) - \alpha\left(\frac{1}{n}\sum_{i=1}^{n}g(x_i) - g_{av}\right)$$

where α is a constant. For large values of n, the term $\frac{1}{n}\Sigma_{i=1}^{n}\,g(x_i)$ will be approximately equal to g_{av} and hence, the bracketed term multiplied by α will be approximately zero. The advantage of this is that, as long as we choose $g(x)$, the control variate, appropriately, we can reduce the variance in the simulation.

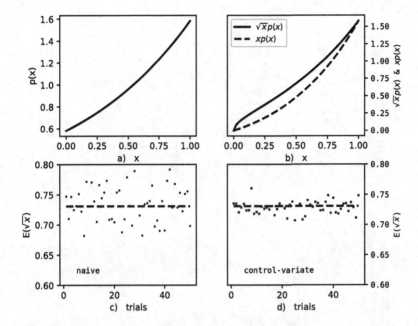

FIGURE B.4 (a) $p(x)$, (b) $\sqrt{x}\,p(x)$ and $xp(x)$, (c) scatter without control variate, (d) scatter with control variate

To illustrate this, we'll calculate the expectation value of $h(x) = \sqrt{x}$ over the range $x = 0$ to $x = 1$, where we'll use the asymmetric probability density function $p(x)$, given by $p(x) = e^x / (e-1)$ (this is plotted in Figure B.4(a). For $g(x)$ we'll choose $g(x) = x$, for which the exact expectation value over the same range is $g_{av} = 1/(e-1)$. Figure B.4(b) shows the comparison between the kernels, $h(x)p(x)$ and $g(x)p(x)$. It is possible to determine an optimal value for the constant α; however, we'll just use the arbitrary value of 1 here. This leaves us with:

$$E\left(\sqrt{x}\right) \approx \frac{1}{n}\sum_{i=1}^{n}\sqrt{x_i} - \left(\frac{1}{n}\sum_{i=1}^{n}x_i - \frac{1}{e-1}\right)$$

We can obtain random numbers, x, from the inverse cumulative distribution function corresponding to $p(x)$ using $x = \ln\left((e-1)z+1\right)$, where z is a uniform random variable in the range $0 \leq z \leq 1$. Figures B.4(c) and B.4(d) show the resulting scatter for the expected value of $\sqrt{x}$, both with (Figure B.4(d)) and without (Figure B.4(c)) the use of the control variate.

Index

Printed in the United States
by Baker & Taylor Publisher Services